Space Exploration

Look for these and other books in the Lucent Overview series:

Abortion
Acid Rain
AIDS
Alcoholism
Animal Rights
The Beginning of Writing
Cancer
The Collapse of the Soviet Union
Dealing with Death
Death Penalty
Democracy
Drugs and Sports
Drug Trafficking
Eating Disorders
Endangered Species
Energy Alternatives
Espionage
Extraterrestrial Life
Gangs
Garbage
The Greenhouse Effect
Gun Control
Hate Groups
Hazardous Waste
The Holocaust
Homeless Children
Illiteracy
Immigration
Ocean Pollution
Oil Spills
The Olympic Games
Organ Transplants
Ozone
Population
Prisons
Rainforests
Recycling
Reunification of Germany
Smoking
Space Exploration
Special Effects in the Movies
Teen Alcoholism
Teen Pregnancy
Teen Suicide
The UFO Challenge
The United Nations

Space Exploration

by Anita Louise McCormick

LUCENT BOOKS

LUCENT Overview Series

LUCENT Overview Series

Library of Congress Cataloging-in-Publication Data

McCormick, Anita Louise.
Space exploration / by Anita Louise McCormick.
p. cm. — (Lucent overview series)
Includes bibliographical references and index.
Summary: Discusses a variety of issues related to space exploration, including the exploration of Mars, the need for a moon base, Space Station Freedom, etc.
ISBN 1-56006-149-9 (alk. paper)
1. Outer space—Exploration—United States—Juvenile literature. 2. Astronautics and state—United States—Juvenile literature. [1. Outer space—Exploration. 2. Astronautics.] I. Title. II. Series.
TL789.8.U5M337 1994
919.9'04—dc20 93-1830
CIP
AC

P.O. Box 289011, San Diego, CA 92198-9011
Printed in the U.S.A.

Contents

Introduction

SPACE RESEARCH IS a booming industry. Many of the nations that once relied on America to launch their communications satellites and space research experiments are now developing their own launch programs. Japan, China, Russia, and the European Space Agency (ESA) all operate satellite-launch programs, and they are becoming increasingly competitive in this field. The ESA, for example, has already cornered about half of the world's commercial satellite-launch contracts. It plans to invest about $4.5 billion in its Ariane-5 launch rockets by the year 2000. Independent U.S. aerospace companies are also starting to provide launch services for satellites and other small research projects.

Satellite launches

Satellite launches are perhaps the most profitable part of the growing field of space research. Since 1957, when the Soviets launched the world's first earth-orbiting satellite, more than 3,500 satellites have been launched into orbit for what amounts to a $1 billion-a-year business. Many technological breakthroughs, in fields ranging from microelectronics to medicine, have resulted from space research. Thus, the field involves a great deal of money and prestige.

(Opposite page) Just seconds after ignition, space shuttle Discovery *rockets toward earth's orbit. Once in orbit, space shuttles like the* Discovery *are often used to deploy satellites.*

The United States long ago earned a reputation

as a leader in space research and technology. But the United States cannot rest on its past accomplishments if it wishes to remain a leader in this increasingly crowded field. Currently the U.S. government spends about $15 billion a year on space research. Although this amount may change, the central issues that the nation must consider will not. Among these are: Who should conduct space research? What aims should this research have? How can these aims best be accomplished?

In its quest to help expand our knowledge of the universe, space shuttle Challenger *soars over a cloud-covered earth.*

NASA

As for who should conduct the research, the National Aeronautics and Space Administration (NASA) has overseen nearly all of America's space research so far. Some industry observers believe NASA remains in the best position to continue that work. It has repeatedly shown its "resiliency, innovation, and skill" in its role as America's premier space agency, says NASA legislative affairs specialist John Lawrence. Others, including writer George Henry Elias, have lost faith in NASA's abilities and believe the agency should be replaced. "NASA, the proud agency that once bore the nation's hopes in space," Elias writes, "has outlived its usefulness. It is an aged and faltering institution whose shortcomings have become embarrassingly evident to all but the most self-interested."

Numerous suggestions have been made over the years for changing NASA's leading role in America's space program. Some have suggested stronger military involvement, whereas others want to strengthen the independence under which NASA has developed. Another, more recent, proposal suggests dividing space research tasks between NASA, private industry, and the military. Under this proposal, NASA would oversee space exploration and crewed missions, private industry would operate routine satellite deployments and handle small payload launches, and the military would control projects in space involving the national defense. Debate on this proposal is likely to continue for some time.

Meanwhile, scientists, politicians, and others wrestle with what direction—among the many possible directions—America's space program should take. There are many disagreements within the scientific community about how space research dollars are best spent—on earth-based or orbiting telescopes, on satellites, on mechanical probes, on crewed space shuttles, on building a permanent space station or moon base, or on researching ways of sending humans to Mars.

Whatever the answers to these questions, America's continued presence in the field of space research seems assured. The Augustine Committee, an advisory group formed in 1990 to examine the U.S. space program, believes this research has become an essential part of our lives. The Augustine Committee writes:

> American scientists and engineers have used opportunities for access to space to advance human understanding of ourselves, our planet, our solar system, and our universe. . . . These achievements and the understanding gained from them will continue to be one of the most significant products of the nation's investment in the . . . space program.

1

Why Explore Space?

THE U.S. SPACE program is one of the most fascinating scientific endeavors of the last half century. Through NASA, the main agency of the U.S. space program, scientists have found ways to do things people once only dreamed about. Methods have been developed to lift heavy objects up through the atmosphere on gigantic rockets; make satellites orbit, or circle, the earth; send scientific probes on voyages to distant planets; and send human beings safely into space. Through the work of the space program, astronomers and other space scientists have learned many fascinating things about the nature of the universe. They have even discovered many previously unknown facts about the earth itself.

The space program, even at its inception, was dogged by questions of its usefulness. Many citizens and government officials did not initially see value in space research and exploration. Cost was a big concern. Running a space program is expensive. It requires research in many technical fields as rocketry, computers, physics, electronics, and aerodynamics. The vast amount of costly research necessary to launch even the simplest of space probes made it impossible for universities,

(Opposite page) Astronaut Bruce McCandless II takes a historic space walk in 1984 to test a device that allows astronauts to move freely in space without being tethered to a space shuttle. Just decades ago, human space exploration was possible only in our dreams.

private institutes, or corporations to raise enough money to launch a rocket on their own. Any space program would require direction and money from the federal government.

Many well-known scientists tried to convince the U.S. government to fund space research, but little was done until the mid-1950s. President Dwight D. Eisenhower felt that the United States should have some sort of a space program, and that it should be under military control. In July 1955, he proclaimed that the United States would launch a satellite into space by 1958. Although President Eisenhower helped to set the goal of a satellite launch for America, he was slow to push Congress for the funding needed to make it possible.

The cold war space race

Congress was reluctant to spend money on space research when so many other programs were in need of funding. For the six years between 1955 and 1961, Congress authorized only $4.5 billion to be used for all space research. It took the Soviet Union's launch of *Sputnik 1*—the world's first artificial satellite—on October 4, 1957, for President Eisenhower and Congress to make a serious commitment to funding a space program.

Dwight D. Eisenhower was the first U.S. president to push for an American space program.

When the Soviet Union launched the satellite *Sputnik 1*, the United States and the Soviet Union were involved in a cold war, which meant that even though there was no actual fighting going on between the two countries, neither government trusted the other. Both the United States and the Soviet Union were world superpowers, and both possessed nuclear weapons. This made it possible for either country to launch a devastating attack on the other. The U.S. government did not want the Soviet Union to surpass America in mil-

The Soviet Union's launch of Sputnik 1, *the world's first artificial satellite, frightened some American leaders. They did not want the Soviet Union to surpass the United States in science and technology.*

itary strength. Leaders were concerned about the possibility of the Soviet Union developing even more advanced weapons systems than they already had.

When the Soviet Union surprised the world by successfully launching the first satellite into orbit, American citizens, as well as their leaders, were frightened. The Soviet Union had suddenly moved ahead of the United States in a very important area of scientific and military accomplishment. Some U.S. officials feared that, with a satellite in orbit, the Soviets would have a new way to spy on the United States—or that, perhaps in the future, Soviet spacecraft could be used to drop bombs on American soil. President Eisenhower, who had hoped that the United States would be the first nation to put an object into space, finally pushed through congressional funding for a U.S. space program.

The new program's first success came on January 31, 1958, when the United States put a satellite, *Explorer 1*, into orbit around earth. *Ex-*

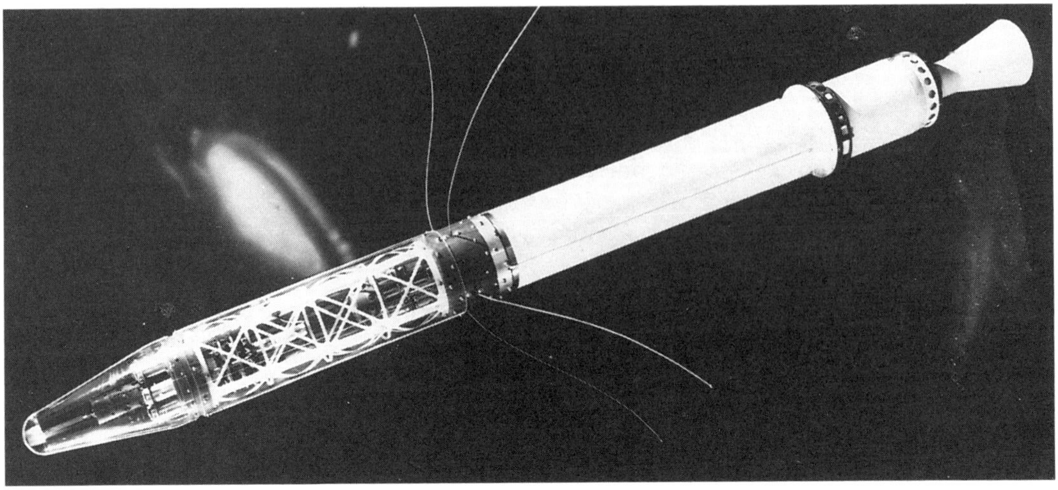

A model of Explorer 1, *the first American satellite launched into orbit. This successful mission helped convince Congress of the value of a full-fledged space program.*

plorer 1 reached a higher orbit than did *Sputnik 1* because it was smaller, weighed less, and was boosted into space by a more powerful rocket. *Explorer 1* had a maximum altitude of 984 miles above earth, whereas *Sputnik 1* only went as high as 588 miles. As a result of its higher orbital path, *Explorer 1* helped scientists make one of the first important discoveries of the space age. The satellite transmitted data back to earth about the doughnut-shaped Van Allen radiation belts that encircle the planet. These belts protect the earth from the sun's high-energy radiation particles by entrapping the particles in its magnetic field.

While the *Explorer 1* mission proved to the world that America could successfully launch a satellite, the nation still did not have a plan for its future in space. For a space program to excel in the long term, it needed funding and direction. To deal with the problem, Congress studied the feasibility and cost of running a full-fledged space program. Astronomers and other scientists were called upon to testify. On July 29, 1958, President Eisenhower signed the National Aeronautics and Space Act, officially establishing NASA.

Eisenhower's science advisor, James Killian,

wanted the country to have a space program, but was concerned that the American-Soviet rivalry would get in the way of true scientific research:

> I believe that in space exploration, as in all other fields we choose to go into, we must never be content to be second best, but I do not believe that this requires us to engage in a prestige race with the Soviets. We should pursue our own objectives in space science, and not let the Soviets choose them for us by copying what they do. We should insist on a space program that is in balance with our other vital objectives and that does not rob them because they are currently less spectacular.

Goals for space

When President John F. Kennedy came into office in 1961 the U.S. space program got a boost. Kennedy favored an aggressive program, one that would surpass Soviet efforts. While Eisenhower and many of his advisors had wanted to concen-

Astronaut John Glenn explains to President John F. Kennedy how astronauts use their gloves in space. Many Americans were inspired by Kennedy's vision of a strong space program and human space flight.

trate on launching orbiting satellites, Kennedy wanted to see the United States put humans in space. He gave the new NASA space program a new sense of importance when he proclaimed that space "is our new great frontier" and challenged the nation to engage the Soviet Union in a space race. He addressed a joint session of Congress on May 25, 1961, saying, "I believe that this nation should commit itself to achieving this goal, before the decade is out, of landing a man on the moon and returning him safely to Earth."

At the time of Kennedy's speech, NASA had very little experience in human space flight. NASA had only launched one person—Alan B. Shepard—into space and that suborbital flight lasted only fifteen minutes. It took Shepard 116.5 miles up into space and concluded with a "splashdown" ocean landing, only 304 miles away from the launch pad.

A military helicopter hoists astronaut Alan B. Shepard to safety after his fifteen-minute suborbital flight ended with a "splashdown" ocean landing.

President Kennedy believed that a fully funded national space program would be an ideal way of pushing American technology forward. He also felt it would provide a common goal for the nation's best scientific minds:

> We choose to go to the moon in this decade and decide to do other things [in space] not because they are easy, but because they are hard, because the goal will serve to organize and measure the best of our energies and skills, because the challenge is one that we are willing to accept, unwilling to postpone, and one we intend to win.

By all estimates, it would be an expensive goal. Scientists estimated that to put an astronaut on the moon in less than ten years would require between $20 billion and $40 billion. At first, many members of Congress were hesitant. They worried about voter reaction to such an expensive program. But when they considered the rousing welcome the public had given America's first astronaut, Alan B. Shepard, when he returned from his mission, they decided to approve the funds needed for the nation's first comprehensive space program. They gave NASA $1.7 billion to start the program.

New views of the universe

Although the U.S. space program had its roots in the American-Soviet rivalry, many scientists viewed the space program as an open door to important new studies of the universe. All astronomical study had been earth-based up to that point, so scientists' view of the universe was severely limited. The gas molecules in the earth's atmosphere distort a telescope's view of stars and planets so they cannot be viewed clearly—even with the most powerful of lenses. The atmosphere also blocks high-frequency radiation such as infrared rays, gamma rays, and X rays. These rays contain vital information astronomers can use to better understand the stars and galaxies. Because of the space program, probes and satellites could be rocketed above the atmosphere, providing scientists with data that had never before been available. By studying the information radioed back to earth from satellites and orbiting probes, scientists could test past theories, develop new ones, and come to a better understanding of the universe.

Many goals have been accomplished through NASA space projects. During the first decade of the space program the United States launched satellites into orbit, sent probes to nearby planets, and safely transported human beings to the moon and back. Since then, NASA has built space shuttles, space stations, orbiting satellite communication networks, and an orbiting telescope. It has also sent probes to distant planets—even beyond the boundaries of the solar system. As a result of these and other projects, scientists have learned things about the universe that they could never have discovered using only earth-based equipment.

Space shuttle and satellite launches have been an important part of NASA's many activities over the past decade.

Worthier causes?

Despite achievements, there is a growing sentiment that money spent on space research could be better used to help solve some of the many serious problems here on earth. Education, poverty, hunger, disease, and even the national deficit are all worthy causes that could benefit from some of NASA's money. Gar Smith, editor of *Earth Island Journal*, said, "We need people programs rather than a space program."

Even though the space program has given many benefits to the scientific community, some scientists worry that attention and resources devoted to the space program take away from attention and resources other important scientific pro-

jects deserve. James R. Killian, former president of the Massachusetts Institute of Technology, said, "Many thoughtful people are apprehensive that our space program with all its drama and challenge will get out of hand and will absorb talent and funds to the detriment of our long-term strength in military and industrial technology and basic science."

Those who favor continued space research argue that the benefits of the space program make it well worth its annual multibillion-dollar budget. They contend that the knowledge gained from space research is useful and important to the human race. Many discoveries made in the process of space research have been adapted to needs on earth. Cornell astronomer Carl Sagan said, "Planetary exploration has many virtues. It permits us

Astronomer Carl Sagan believes that the benefits of space research to the sciences, and to such industries as communications and electronics, justify the costs.

to refine insights derived from such Earth-bound sciences as meteorology, climatology, geology and biology, to broaden their powers to improve their practical applications here on earth."

Watches and microwave ovens

The electronics industry, for example, has benefited tremendously from research conducted by NASA scientists. The lightweight, miniaturized electronic parts needed for satellite and space shuttle computers and other equipment have been adapted for use in many everyday items such as calculators, computers, and digital watches. Before NASA researchers made advances in microelectronics, computers were bulky, difficult to use, and much too expensive for anyone but well-funded research institutes to afford. Now, they are commonly used in libraries, businesses, homes, and schools.

The space program involves many different kinds of research. In addition to developing the technology needed to lift rockets off the ground and keep them functioning, NASA studies ways to feed, clothe, and shelter astronauts sent on space missions. Small microwave ovens, similar to those now used in homes, were first developed for use by astronauts in space. The protective materials commonly used in firefighters' clothing were first developed to shield NASA astronauts from extreme temperatures. The background research for many high-technology medical devices found in hospitals, including patient-monitoring equipment used in intensive care, operating rooms, and trauma units, came from the space program. These, as well as numerous other products that originated in the space program, have made their way into general use.

NASA's work in microelectronics has led to compact, easy-to-use, inexpensive computers.

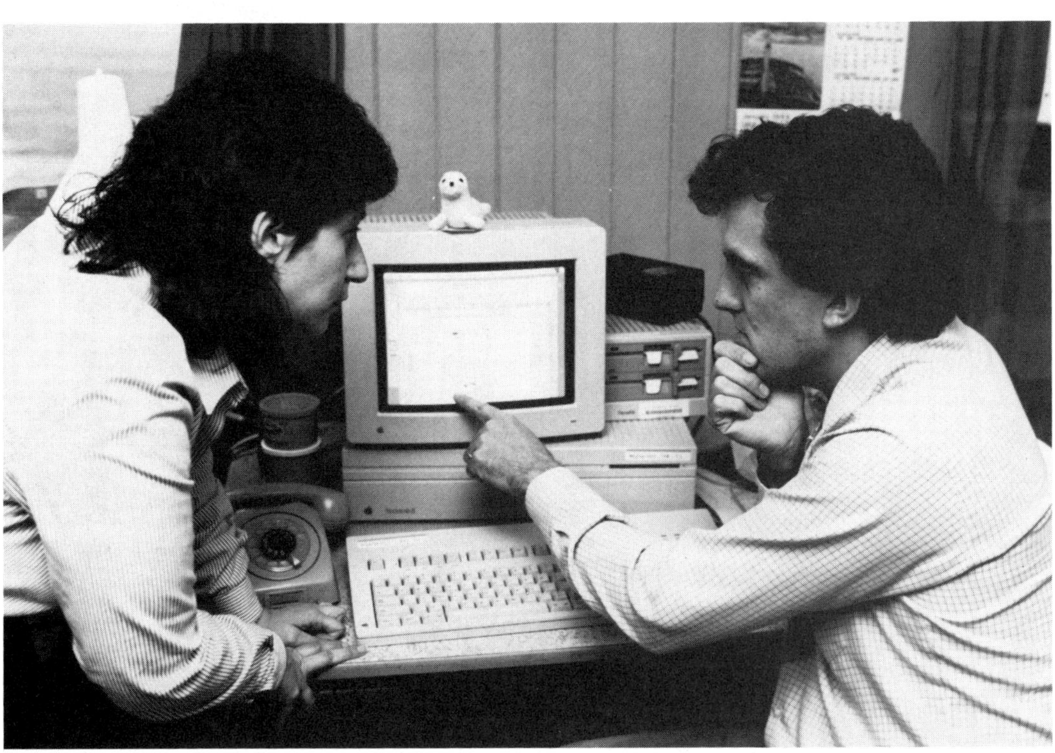

Satellites are another example of how NASA research has benefited people on earth, and they are one of the most commercially successful products of the space program. They make reliable worldwide communications possible. Before satellites were launched into orbit, international telecommunications could only be transmitted by two imperfect methods: through underwater telephone cables or through the airwaves via shortwave radio. Cables can only handle a limited number of calls at one time, and shortwave radio communication is often unreliable due to changing atmospheric conditions.

With satellites, network television and radio programs, international telephone calls, and many other communications can be instantly retrans-

A communications satellite in orbit over Zaire, Africa. Satellites provide quick transmission of television, radio, and telephone signals over continents and oceans.

mitted to domestic or overseas locations. A network of three satellites can cover all of the world except the extreme north and south polar regions. Cable television companies use satellite technology to receive most of the programming they provide to their viewers. People who own backyard satellite dishes can view many of these programs directly.

Broadcasters are not the only users of satellite communications. Many newspapers use satellite technology to receive up-to-date information from their field correspondents and from national news services. Some newspapers, such as *USA Today* and the *Wall Street Journal*, use satellites to transmit the contents of their paper to printers in various parts of the country. This allows them to distribute papers much faster than they could if they used only one printing facility. Many businesses use satellites to relay information on sales, profits, and other developments to offices around the world. Stock market news is transmitted over satellites. Satellites are also used to help rescue crews pinpoint the location of an airplane's black-box emergency transmitter if the plane crashes in a remote location.

Observing the earth's atmosphere

In addition to transmitting global communications, satellites provide scientists with fascinating new views of earth and the universe. NASA has placed satellites and probes into orbit that will collect information about the earth and radio it back to research stations around the world. Satellites make it possible to observe weather systems from space, giving weather forecasters more accurate information and pictures of the atmosphere on which to base their forecasts. When forecasters can give people advanced warning of hurricanes and other tropical storms, the people have enough time to board up their homes and seek shelter before the severe storms hit, and many lives are saved.

A new generation of earth-observing satellites is helping scientists keep a closer watch on environmental problems. By placing sensitive cameras and monitoring equipment on satellites above the earth's atmosphere, scientists can gather data that would be difficult, if not impossible, to obtain from the ground. The equipment can detect and watch oil spills, forest fires, acid rain damage, water pollution, the hole in the ozone layer, and many other environmental disturbances.

An artist illustrates one of a new generation of earth-observing satellites. This satellite is designed to gather data on the upper atmosphere in order to better understand and predict changes in climate.

Satellites and national defense

Satellites also play an important role in America's national defense system. The military uses satellites to keep a close watch on the activities of hostile governments. Nuclear weapons tests,

The first launch of the Trident missile at Cape Canaveral in Florida in 1978. Today, satellite technology has many military uses, including the monitoring of missile launches and nuclear weapons tests.

rocket launches, weapons movement, and other potentially hostile activities in foreign countries can now be closely monitored with advanced satellite technology. During the Persian Gulf War in 1991 and 1992, satellites were used to detect Iraqi Scud missiles launched toward Israel and Saudi Arabia. This enabled American and allied troops to launch Patriot missiles that intercepted and destroyed most of the Scud missiles before they could reach their targets. Satellites were also used to monitor Iraqi troop movement.

The space program has existed for little more than thirty years now. But in that short period of time, it has revolutionized the way human beings communicate, study the universe, and observe the earth in peace and war.

2

Exploring Space with Telescopes

THE IMMENSE DISTANCES between the planets and stars make it impossible for astronomers to visit—or even obtain samples from—the vast majority of celestial objects they want to study. Because of this, nearly all space research has to be carried out indirectly. Scientists must learn about the universe by analyzing the radiation generated by or reflected from objects in space. Telescopes are some of the most important tools of astronomical research, because the radiation they collect supplies scientists with many kinds of information.

Light-gathering optical telescopes have long been used by astronomers to study the moon, stars, and planets. But light-wave radiation represents only a small portion of the energy generated or reflected by celestial objects. Other kinds of waves—radio waves, infrared waves, X rays, and gamma rays—also transmit information. Special kinds of telescopes are needed for scientists to receive the data that is constantly being transmitted on these high-frequency wavelengths. In recent years, orbiting optical telescopes and orbiting high-frequency telescopes have given astronomers much new information.

(Opposite page) Telescopes of all types, including these radio telescopes in New Mexico, allow astronomers to learn more about distant stars and planets barely visible to the naked eye.

There are some disagreements within the scientific community over which kinds of scientifically advanced telescopes are most worthy of funding. Each type of telescope has advantages as well as drawbacks.

Space telescopes

Space telescopes—telescopes that are literally out in space—have the advantage of magnifying the universe without the distorting effects of the atmosphere. This makes it possible for astronomers to view known objects more clearly than they have in the past and to observe distant stars and galaxies that have never been charted.

Space telescopes can also gather data on wavelengths that cannot pass through the earth's atmosphere. Robert Jastrow, an astronomer and the founder of NASA's Goddard Institute of Space Studies, said:

> Some of the most interesting objects in the Universe, such as black holes and quasars, emit copious amounts of just those kinds of radiation that cannot get to the ground—particularly gamma rays and X rays. And stars with planets forming around them emit copious amounts of infrared radiation, another kind of radiation blocked by the atmosphere. . . . Most of the information about these exotic, important objects was denied to astronomers before the advent of the space age.

The Hubble Space Telescope has provided scientists with a view of galaxies estimated to be 7 billion to 10 billion light-years away.

In November 1992, the orbiting Hubble Space Telescope gave scientists an unparalleled view of what could be the most distant cluster of galaxies ever sighted. Astronomers estimate these galaxies to be 7 billion to 10 billion light-years away. The light the Hubble telescope observed from these galaxies has been traveling through the universe for several billion years. So instead of seeing these galaxies as they appear today, astronomers have a picture of how they looked in their early stages of formation, shortly after the universe formed. Alan

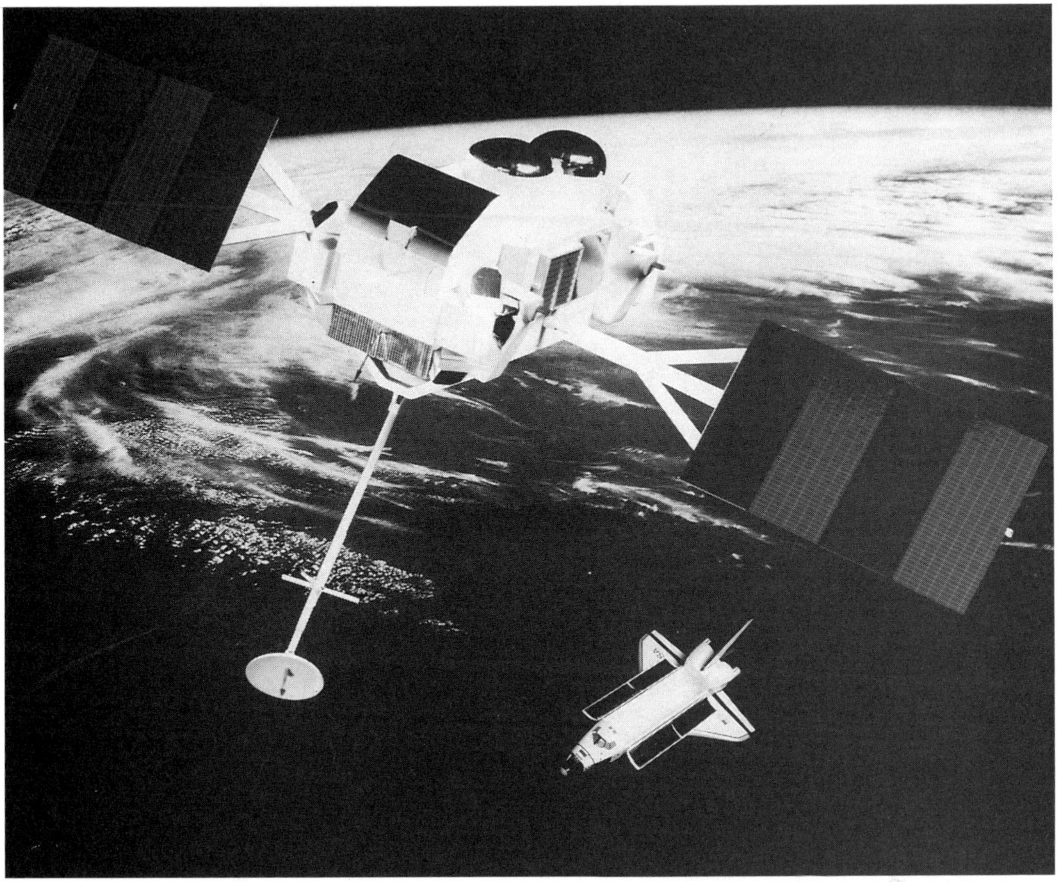

An illustration of the orbiting Gamma Ray Observatory, whose telescopes gather data on gamma radiation, pulsars, quasars, and black holes.

Dressler, an astronomer with the Carnegie Institution in Washington, D.C., said that if these galaxies could be viewed with an earth-based telescope, they would "look just like a bunch of fuzzy blobs" because of distortion caused by gas molecules in the atmosphere.

The Hubble has helped to make many other important scientific breakthroughs possible. It helped scientists confirm the theory that the sun may not be the only star to have a family of planets. In late 1992, the Hubble gave astronomers their first view of a distant star circled by rings of orbiting matter. "We have found a place where it is very possible that there will be planets within the next few million years," said Edwin J. Weiler, Hubble project scientist. "This takes us closer to the final proof that there are other planets where there could be life."

Problems with the Hubble

The Hubble Space Telescope has helped scientists discover many exciting things about the universe, but NASA could have a difficult time convincing Congress to approve funding for any similar space telescope projects. The Hubble cost about $1.5 billion, which is higher than the combined cost of all the giant earth-based telescopes. Technical problems and disagreements between NASA management and project contractors nearly doubled the final cost of the Hubble. The biggest blow came after the Hubble's April 1990 launch, when NASA discovered that the telescope's primary mirror had been precisely ground

Space shuttle Discovery *releases the $1.5 billion Hubble Space Telescope into orbit in 1990.*

to the wrong specifications. The Hubble was supposed to detect large planets orbiting nearby stars, but so far the faulty mirror has prevented it from clearly seeing these planets.

These setbacks have been extremely frustrating for space researchers who have waited anxiously for years for the Hubble to go into service. Science writer Eric J. Lerner said:

> The flaw has become a symbol to the public of what is wrong with NASA, the American space effort, and big science generally. Scientists and lay people alike wonder how a mirror reputed to be the most precise in the world, accurate to 1/64th of a wavelength of light, could turn out to be in error by 30 times as much.

Space shuttle astronauts are scheduled to go to the Hubble to repair and modify it within the next few years. The first such mission has been scheduled for December 1993. Astronauts will make repairs during several spacewalks. Missions such as this are costly, risky, and time consuming. The repair will cost NASA about $30 million, and success is not guaranteed. Spacewalk repair missions involving malfunctioning satellites and other NASA equipment have a mixed record of success. Although astronauts take many precautions during repair missions, the risk of injury or death is always greatest when astronauts leave the controlled environment of their spacecraft. There is some disagreement over whether this type of risk is worth the gains.

Another drawback of space telescopes is that they can be easily damaged if they are hit by a meteorite or by a piece of earth-generated space junk. Damage of this sort would put the Hubble out of service and make costly repairs necessary. Dr. Robert Jastrow said:

> There are more than 7,000 pieces of man-made debris—dead satellites, empty rocket casings, and fragments of destroyed space craft—that are orbiting the Earth and big enough to be tracked by radar. . . . A collision with any of these thousands of objects would probably destroy the Hubble.

Stung by criticism over the Hubble's problems and costs, scientists are looking at ideas for less expensive, special-purpose space telescopes. An $800 million orbiting Space Infrared Telescope now being planned by NASA will help scientists gather information about the universe on infrared wavelengths. Because infrared waves cannot pass through the earth's atmosphere, such research can be accomplished only with a space telescope.

Light-gathering telescopes

Although the Hubble's above-the-atmosphere view of the universe offers many advantages, some astronomers believe that highly sophisticated earth-based telescopes, such as the Keck Telescope in Hawaii, offer more to the scientific community for less cost. The Keck Telescope gathers both visible light and near-infrared radiation, which is just below the wavelength of visible light. Its design is far more advanced than any light-gathering earth-based telescope built to date. The Keck's huge main mirror is 400 inches wide—nearly double the size of any other light-gathering telescope. The main mirror is made up of thirty-six individual hexagonal segments, which are adjusted by computer to keep objects in clear focus. The Keck Telescope also uses the latest technology to record and analyze the astronomical data it collects. The Keck changes incoming light from stars or planets into electronic signals. The data gathered by the Keck can then be stored, analyzed, displayed as a picture on a computer screen, or transmitted by computer modem to astronomers anywhere in the world.

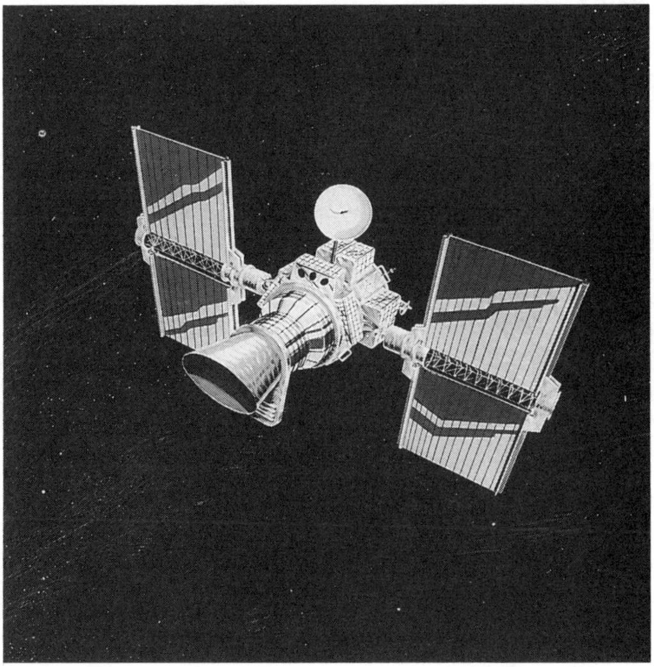

The Space Infrared Telescope will gather information about the universe on infrared wavelengths, which cannot pass through the earth's atmosphere.

While the Keck has certain advantages over the Hubble Telescope, there are also disadvantages. Like all earth-based optical telescopes, the Keck's view of the universe is hampered by earth's atmosphere. Sky glow, which is caused by artificial lighting in cities, towns, and highways, also hampers an earth-based optical telescope's ability to view the universe. To minimize these problems, high-quality telescopes are usually built at remote high-altitude mountain locations, where the air is much thinner than it is at sea level. The Keck Telescope is located at one of the best observation sites in the world—the summit of Mauna Kea in Hawaii, which is 4,600 feet above sea level. At this height, it can be used to observe both light waves and near-infrared radiation. Even at this height, there is still a slight problem with atmospheric distortion, but scien-

Located on the summit of Mauna Kea in Hawaii, the Keck Telescope is the most advanced of all light-gathering optical telescopes.

tists are working to develop a system that will use computers to compensate for air movement and changes in temperature and humidity. Once this computer system is completed, it will improve the clearness and stability of images the Keck Telescope receives.

A second Keck Telescope

The Keck Telescope cost about $70 million to design and build. It is the most expensive optical telescope on earth. Still, it was far less expensive than the Hubble, and most of the money came from the privately funded Keck Foundation. The Keck Foundation also donated $74.6 million for a second, identical telescope to be built next to the first, and it is now under construction.

Earth-based telescopes such as the Keck can be modified and repaired far more easily than space-based telescopes. Because of this, some astronomers believe that they are a better use of money than orbiting telescopes. Improvements can be made without costly and risky space shuttle missions.

When the second Keck Telescope is completed, it will be used in conjunction with the first to make even more detailed observations of the universe. The director of NASA's Jet Propulsion Laboratory, Edward Stone, said, "The construction of a second Keck Telescope will result in an observatory with an unparalleled combination of light-gathering power." By combining the images from the two Keck telescopes with images from four smaller (60-inch lens) telescopes and connecting their incoming signals with computers, a network will be created that has the resolution of a 2,755-inch lens. Until recently, scientists could only speculate about whether or not other stars had planetary systems. When the new observatory is completed, astronomers predict they will

Radio telescopes have been responsible for many important discoveries, including the identification of hundreds of previously uncharted quasars and pulsars.

be able to see Jupiter-sized planets orbiting stars as far as 50 light-years away.

Radio telescopes

Radio telescopes are another important way for space researchers to gather data. They are extremely valuable because they are able to receive information at energy wavelengths not accessible to light-gathering telescopes. Radio telescopes can detect radio signals in the cold regions of gas that give birth to stars and in hot regions of gas associated with exploding stars and galaxies. Radio signals also travel through the heavy dust cloud regions of space that block the view of light-gathering telescopes. In the past few decades, many important discoveries have been made by astronomers using radio telescopes. Hundreds of previously uncharted quasars (radio stars) and pulsars (pulsating radio stars) have been identified.

Astronomers also need radio telescopes to study objects within the solar system. Radio telescopes were used to determine Venus's speed of rotation. Because Venus has such a heavy cloud cover, astronomers were not able to view the planet's surface with light-gathering telescopes. Radio waves, however, could pass through the clouds. By bouncing radio signals off Venus and analyzing the returning echo signal, astronomers could see Venus's surface. From this they determined that Venus rotates once every 243 earth days. NASA recently used its 230-foot radio antenna dish in the Mojave Desert of California to take pictures of the asteroid Toutatis. Signals were bounced off the asteroid as it passed within 2.2 million miles of the earth in late 1992. The returning radar signals were then assembled into pictures by computer. "This is our first clear look at one of the many asteroids whose orbits can in-

Radio telescopes have provided astronomers with a previously unseen view of the surface of Venus.

tersect Earth's orbit," NASA astronomer Steven Ostro said. "The pictures of Toutatis are 100 times more detailed than the best previous images of a 'near-Earth' asteroid."

SETI and radio signals

Radio telescopes are also being used in the search for extraterrestrial intelligence (SETI). Many people feel SETI is one of the most exciting uses of radio astronomy. It could help to answer the question "are we alone in the galaxy?" Scientists around the world have been searching for extraterrestrial signals since the early 1960s. So far, they have not been able to receive a signal that they are sure is from an extraterrestrial source. Broadcasts from radio and television stations here on earth, as well as signals relayed by satellites and other craft, make it increasingly difficult for astronomers to distinguish possible extraterrestrial transmissions from those generated by humans.

The Planetary Society, one of the largest astronomy groups in the United States, has worked with Harvard University to build a small, privately funded SETI program. Australia, Canada, France, Germany, Holland, and Russia have also conducted SETI research in the past few decades. The U.S. government, through NASA, is taking part in this search.

NASA's newest SETI system (the High-Resolution Microwave Survey, or HRMS) was switched on on October 12, 1992. It uses some of the world's largest radio telescope dishes—including the 112-foot-wide antenna dish in Goldstone, California; the 1,000-foot-wide dish at Arcibo, Puerto Rico; and several other radio telescope facilities. Advanced computer networks are being used to monitor billions of possible communications frequencies and analyze the signals

Are we alone in the galaxy? NASA is attempting to answer this question with radio telescopes that search in space for signals from extraterrestrial beings.

received. In order to find any intelligent transmission, computers will sort through not only signals from earth, but also signals generated by natural phenomena in the universe such as pulsars.

NASA's most advanced telescopes

The new SETI project is the most advanced radio telescope system NASA has ever constructed. NASA has already spent nearly $40 million on the SETI program, and the total cost of the project by the end of the decade is estimated to be about $100 million. Between 1992 and 2001, NASA will scan 10 billion times more territory with the SETI radio telescopes than it has in all previous surveys combined. At any one time, 10 million different frequencies can be monitored. The antenna-computer systems will analyze frequencies and sensitivity levels that were not possible for astronomers to monitor in the past. "We are just at the beginning of such searches," said Carl Sagan. If signals from intelligent life are found, "We will have transformed our civilization and our concept of ourselves forever."

Reprinted by permission of *Skeptical Inquirer.*

It was not easy for SETI pioneers such as Sagan to convince their fellow scientists and Congress that the program was worth funding. Many astronomers feared that SETI funding would take money away from other projects. Astronomer Charles Seeger said:

> When we found out that the astronomical community was worried that any money spent on SETI would come out of the astronomical research budget, several of us set out to disabuse that point of view; in a friendly way, to educate them that the money would never have come to astronomy if we didn't get it; that SETI was, first of all, astronomy. It took three or four years to persuade the astronomical community that we were a serious, legitimate exercise in hard science.

People in NASA had to fight tough congressional opponents such as Senator William Proxmire, who felt the program was a waste of money, for years before they were able to obtain adequate funding for SETI research. "I think politicians are afraid to take a position which is (a) unconventional and (b) cannot be proven to be right," Bernard Oliver, NASA SETI program chief, said. "It opens them to criticism by people like Proxmire." In the early 1980s, after two prominent scientific committees endorsed the search, the U.S. government finally decided to allocate $1.5 million a year for a SETI program.

If no extraterrestrial messages are received during NASA's ten-year search, critics of the SETI program are certain to insist that the money could have been put to better use. But scientists who support the program contend that whether extraterrestrial transmissions are found or not, the current search is extremely valuable. It is giving astronomers more data about the universe in the microwave radio frequency ranges than they have ever had in the past. "We are surveying the sky at microwave frequencies more thoroughly than anyone ever has," said Tom McDonough, the Planetary Society's SETI coordinator. "By looking for signals from other civilizations, you discover other objects in the universe, like pulsars, for example."

3

Should Humans Be Sent into Space?

(Opposite page) Apollo 11 *astronaut Edwin E. Aldrin Jr. takes historic steps on the surface of the moon. Whether humans should continue to be sent into space remains a question of considerable debate.*

HUMANS HAVE ALWAYS explored. Exploration of new territories is humanity's way of expanding its presence and extending its environment. From ancient times until the present, humans have ventured out of their immediate surroundings to see what lay beyond. Early humans walked across vast plains in search of food and milder climates. Seagoing explorers built exquisitely crafted ships to open new trade routes and satisfy their boundless curiosity about who and what they might find on the other side of the ocean. Pioneers and land-based explorers journeyed over mountains and followed rivers to see where they led and to find prosperity for themselves and their families. Humans even built craft to take them beneath the ocean to see what life existed there.

It seems only natural that humans have also built extraordinary vessels capable of traveling beyond the earth's atmosphere. With these vessels scientists have sought to learn more about the moon, planets, stars, and other phenomena of the universe. No other creatures on earth have the insight, intellect, curiosity, and physical and mechanical capability to explore the unknowns of space and then find ways to make use of the many discoveries. Astronaut Dale Gardner, from NASA, said, "We are meant to travel away from Earth because we have been given the curiosity, the intelligence and the will to devise the means and the wonderful machines to permit such adventures."

Limits of space travel

These are the qualities that inspired the first human space flight—the 1961 journey of Russian cosmonaut Yuri Gagarin—and continue to inspire space exploration today. Space exploration offers limitless opportunities for learning. However, there are practical limits on when and how such learning can be done. These limits play a large role in the debate over sending human explorers into space.

Money is a big factor, especially during difficult economic times when many worthy projects must compete for limited federal tax dollars. Any space exploration is expensive. It requires years of research and years of work developing and testing specialized equipment. Human space flight is especially costly, and most of the things that make crewed missions expensive cannot be eliminated. For astronauts to travel in space, an artificial earthlike environment has to be created. Food, water, and air must be provided. The temperature must be kept within the range of human tolerance, and astronauts must be shielded from deadly radiation. The length of missions is limited by how long an artificial environment can be sustained, and all missions involving humans must include provisions for returning the crew safely to earth.

Uncrewed missions can almost always be ac-

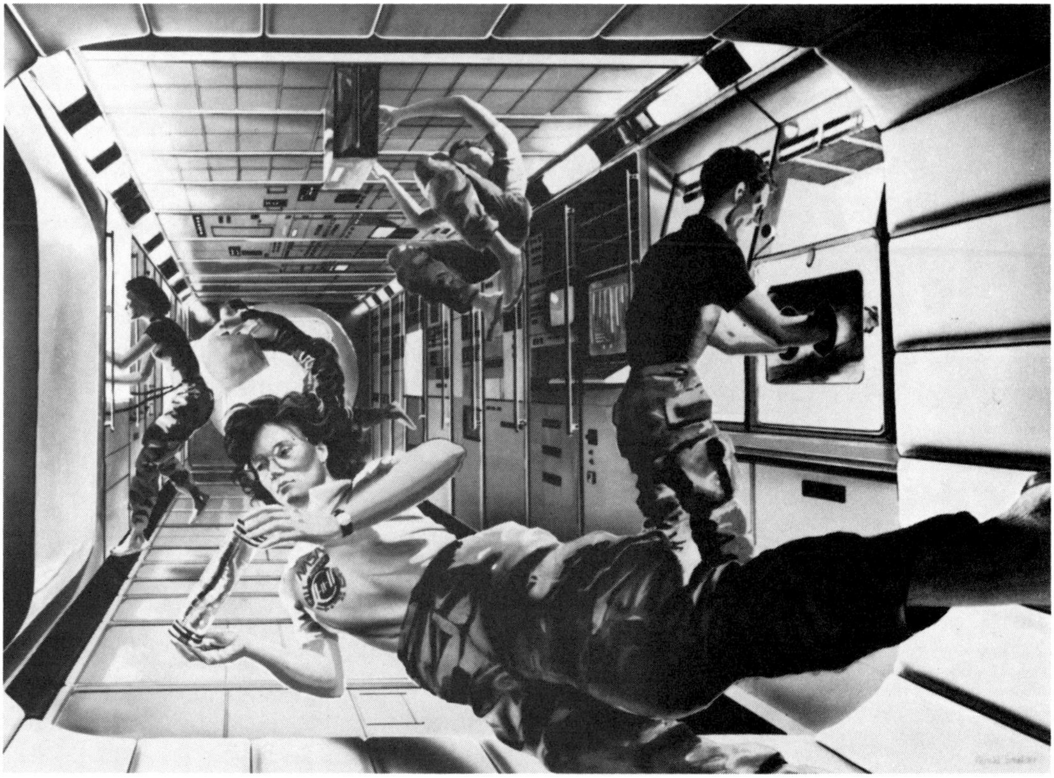

Humans require an artificial earthlike environment for space flight. This need makes crewed missions extremely costly.

complished more economically than those involving humans. No environment needs to be provided. Uncrewed missions can take as long as necessary—weeks, months, or even years. And, because the information collected by probes or spacecraft is relayed back to earth by way of radio, no return is necessary—except in cases where soil samples from another planet or the moon are brought back for scientists to study.

Discoveries without humans aboard

The differences in flight requirements, and thus cost, lead some people to conclude that it makes more sense, especially when money is tight, to explore space using uncrewed spacecraft. Uncrewed spacecraft have played an important role in gathering information about the universe.

The uncrewed *Cosmic Background Explorer* (COBE), for example, gave scientists the data they needed to support the "big bang theory" of how the universe began. With its sensitive monitoring equipment, the earth-orbiting COBE was able to detect faint traces of the radiation that was released from the massive explosion over 15 billion years ago. Space scientists collected data from the COBE for several months and analyzed it by computer for over a year. The information obtained by the COBE permitted them to create a map of galaxies in their earliest stages of formation. David Spergel, an astrophysicist with Princeton University, declared that the probe helped scientists to make "the most important discovery in cosmology in the past 20 years."

The *Pioneer Venus Orbiter*, an uncrewed probe that was in orbit around Venus for fourteen years, has also played an important role in space research. It recently transmitted data back to earth

The Cosmic Background Explorer *has given scientists valuable information supporting the "big bang theory" of the formation of the universe.*

indicating that Venus, which now has a surface temperature of about 900°F, was once a moist, temperate planet that could have supported some primitive forms of life. The data, which were transmitted in October 1992, just before the probe's decaying orbit caused it to burn up in Venus's atmosphere, indicates that Venus may have once had oceans 25 to 75 feet deep and temperatures in the 70°F to 80°F range. Thomas Donahue, of the University of Michigan, said that most scientists now believe that Venus's oceans "vaporized and blew off 3 billion years ago in a runaway greenhouse effect, when the cool early sun increased its luminosity and heated the planet."

Data transmitted by the Pioneer Venus Orbiter *indicate that Venus was once moist and temperate, and could have supported some primitive forms of life.*

Achievements such as these, in addition to the lower cost, lead some space program watchers to conclude that money for exploring space is better spent on uncrewed craft. In the words of science writer Alex Roland, "Most of the payoff in space has come not from manned vehicles but from unmanned. NASA has landed on Mars, visited all the planets in the solar system, and sent probes into deep space all without people on board." Roland is among those who think human space exploration is a high-priced show staged primarily to attract media attention, and that media attention is aimed at keeping public pressure on Congress to fund NASA projects. He characterizes NASA's crewed space program as a "feel-good adventure the country can ill afford."

In this televised view, astronaut Pierre Thuot tries to capture the INTELSAT 6 *satellite while attached to the robotic arm of space shuttle* Endeavour.

A dramatic, human rescue

Most scientists agree that uncrewed probes have made valuable contributions to what is

known about the universe. But many of these scientists also believe that astronauts should remain an important element of the nation's space program. Astronauts have shown their importance in space missions again and again. Robots and probes cannot do everything. Human ingenuity and reasoning powers are superior to machines when it comes to solving problems, adapting to unexpected situations, dealing with crises, or even modifying an experiment when it does not work out as planned.

Astronauts are often able to repair equipment when it breaks down during a space mission. On May 13, 1992, three space shuttle astronauts were able to retrieve and repair the $131 million *INTELSAT 6* satellite during a space walk that lasted a record 8 hours and 29 minutes. The astronauts tried repeatedly to capture the satellite with a 15-foot-long mechanical arm. Although it had worked during rehearsals on earth, the robotic arm did not function properly in the gravity-free environment of space. The astronauts finally resorted to grabbing the satellite with their hands. After many attempts, they caught the *INTELSAT 6*, replaced the faulty rocket booster, and sent it up into its proper orbit. The International Telecommunications Satellite Organization, which owned the satellite, paid NASA $93 million for the rescue effort. Pierre Thuot, one of the astronauts involved in the rescue, said, "The *INTELSAT* mission was a very visual demonstration of humans against machines. The conflict and the drama attracted worldwide attention, but extraordinary things are done on every flight."

Extending the human presence

Scientists who support crewed space flight also argue that if humans are ever to colonize the moon or another planet, such as Mars, they must learn more about how to live and work in a space environment. Technology has reached a point where astronauts *can* live and work in space for extended periods of time. Thus, Americans must decide if they want to spend their money on missions aimed at colonizing space.

Carl Sagan is a strong supporter of human space exploration. He believes that expanding the human presence in the solar system should be a vital part of the space program: "For the first time in human history, it is within the power of this generation to extend the human presence to the other worlds of the solar system—with awe for their wonders, and a thirst for what they have to teach us."

Space shuttles, used to launch nearly half of all satellites, probes, and space telescopes, have played a central role in the space program.

The space shuttle program

The space shuttle program has been the centerpiece of NASA's crewed space program since the early 1980s. It was developed to serve several of the space program's needs. First, shuttles were to provide a place for astronauts to perform scientific experiments on plants, animals, and materials such as metals and crystals under low-gravity conditions. Second, they were to help answer the question of how the astronauts themselves respond to the conditions they must endure on space missions. Medical experiments performed on astronauts aboard the space shuttle test how the human body is affected by living in a near-zero gravity, or microgravity, environment. Science writer A. Royce Dalby praised the space shuttle program: "Until the West has a space station flying, it [the space shuttle] provides the only laboratory available for crucial microgravity research. The program has been and will continue to be a valuable resource for excellent science."

The space shuttle was also designed as an observation platform for scientists to study the earth

The crew of the space shuttle Challenger *included six astronauts and a teacher. Christa McAuliffe (top row, second from left) was to be the first teacher to journey into space.*

from space, and, finally, as a transport vehicle for carrying both astronauts and supplies to and from a future space station. Space shuttles are responsible for launching nearly 50 percent of all the satellites, probes, space telescopes, and other objects that NASA has put into space. And space shuttles are the only NASA spacecraft that can retrieve malfunctioning satellites and, if necessary, return them to earth.

The *Challenger* tragedy

For a long time, the space shuttle's accomplishments have kept the program in high esteem with both the public and Congress. But events in recent years have opened the program and NASA to much criticism and concern. Some believe that the push to add more and more flights to keep public enthusiasm and congressional support high

With millions of television viewers watching, the Challenger *exploded into a flaming ball of fire just seconds after launch.*

has led NASA to become careless. The tragic explosion of the space shuttle *Challenger* has been cited by some observers as an example of problems at NASA.

On the morning of January 28, 1986, well-wishers filled the spectator bleachers at the Kennedy Space Center at Cape Canaveral, Florida. They awaited the flight of the space shuttle *Challenger*. Interest in this flight was high, for the shuttle was to carry the first teacher in space, Christa McAuliffe, along with the shuttle's crew of six astronauts. As friends, families, NASA employees, and millions of television viewers watched, the shuttle was launched. Seventy-three seconds into its ascent, tragedy struck. The shuttle exploded in a flaming ball of fire, leaving behind a horrifying trail of smoke and vapor. All seven people on board died.

Faulty O-rings

Investigations by NASA and an independent commission created by President Ronald Reagan revealed that the explosion had been caused by the failure of the shuttle's O-rings. The O-rings sealed the joints connecting four separate sections of the shuttle's solid rocket boosters. In one solid rocket booster, investigators found that primary and secondary O-rings had failed to form airtight seals. The problem appeared to be the weather. The temperature in Florida on the morning of the launch was unusually low. Investigators concluded that the O-rings were so stiff from the cold that they probably were not flexible enough to form an airtight seal. When the hot gases burned through the primary O-ring, the secondary ring was not in the proper position to stop the gases from burning right through the booster's outer casing.

Investigators discovered that NASA had known about the O-rings' weather-related weaknesses. The engineers who designed the O-rings had warned NASA about launching in such cold temperatures. Yet NASA went ahead with the shuttle launch after concluding that the backup O-rings would correct any possible flaw.

NASA blamed for tragedy

In its final report to the president the Rogers Commission placed responsibility for the shuttle accident with NASA. The commission concluded that NASA failed "to understand and respond to facts obtained during testing." The commission also criticized NASA officials for not halting shuttle flights until the problems had been solved. NASA officials received much of the blame for creating an atmosphere in which managers were afraid to inform anyone that they had a problem even when they suspected one existed. Science writer Neil McAleer said, "We soon all realized that the technical flaws, combined with a lot of minor oversights, poor communication, busy launch schedule pressures, and bad decision-making contributed to the disaster."

Developing new launch systems

The *Challenger* tragedy also revealed another important weakness in NASA's space program. Because NASA had put so much emphasis on shuttles and other crewed flights, it had not given adequate attention to developing other launch systems. When NASA was forced to ground the remaining three space shuttle craft for nearly four years during the *Challenger* investigation, launches of important satellites, space probes, scientific payloads, and even the Hubble Space Telescope were delayed.

Many space researchers would like to see

NASA either increase its fleet of uncrewed rockets or develop a crewed vehicle that is less expensive to fly than the space shuttle. The National Aerospace Plane (NASP), which was first proposed by NASA engineers in 1985, is one alternative to the space shuttle. Instead of having to use huge rocket boosters and a launch pad to get into orbit, the NASP could take off and land on a jet runway. NASA officials predict that, if built, the NASP will make it possible to launch payloads into orbit for less than $4.4 million per flight, as opposed to the $275 million it costs to launch a space shuttle, or the $150 million required to launch an uncrewed Titan rocket. The proposed NASP will have a much smaller cargo bay than the space shuttle, but it will still be large enough to transport most satellites, probes, and other scientific payloads into orbit.

Many people believe that NASA can explore space most effectively by striving for a balance between flights with and without human crews, using each form of exploration to its best advantage. As science writer Robert A. Roe expressed it, "Our [unmanned] space science program was not designed to replace our manned space program. It was designed to be a companion to manned space efforts. Together the two form the exploration and the development balances needed for continuous space expansion."

NASA
USA

4

Should Space Station *Freedom* Be Built?

NASA OFFICIALS AND independent space scientists have worked for nearly a decade toward launching a permanently crewed earth-orbiting space station. The space station, called *Freedom*, was designed by NASA engineers in the mid-1980s. It was to be a huge and sophisticated structure, much larger and more advanced than *Skylab*, NASA's only other space station project.

Skylab, which was launched by NASA in the mid-1970s, was a small, relatively low-technology space station. Leftover hardware from the Apollo program was used for its main components. It was occupied by three crews of astronauts between May 25, 1973, and February 8, 1974. Funding cutbacks left NASA without enough money to rocket astronauts up to the space station, and so it was abandoned. *Skylab*'s orbit gradually decayed until, on July 11, 1979, it fell through the atmosphere and landed in the ocean near Australia.

Freedom was designed to be large and able to serve many purposes. Once in orbit, *Freedom*

(Opposite page) An artist's conception of space station Freedom, *a proposed multipurpose international space laboratory. Due to the huge costs involved, many people wonder whether* Freedom *should be built at all.*

Space station Skylab *in orbit above earth. Lacking the funds needed to rocket astronauts up to the space station,* Skylab *was abandoned after just one year of use.*

was to serve as a research base, a satellite repair and fueling facility, and a stopping-off point for astronauts on missions to the moon or Mars. In the original plans, *Freedom* was over 500 feet long. It was to include a 44-foot-long laboratory and house three trailer-sized habitation modules for the crew to live in. Its designers envisioned space station *Freedom* as an international space laboratory that could eventually house as many as twenty-four scientists and technicians. The addition of a large rectangular structure, known as a *dual keel design*, was planned to provide extra room for more experiments and payloads.

Space station *Freedom* had strong support from President Ronald Reagan, who saw it as an opportunity for America to make advancements in the fields of science and technology. In his 1984 State of the Union address, Reagan said, "A space station will permit quantum leaps in our research in science, communications, and in metals and lifesaving medicines which can be manufactured only in space."

Rising costs

At the time, Reagan estimated that the station would cost $8 billion, and could be launched within a decade. However, by the late 1980s, it became clear that NASA had been too optimistic about plans for space station *Freedom*. Fewer industrial and scientific users than expected had shown interest in the space station, and cost estimates were climbing rapidly. Even after trimming the project several times its cost grew to $38 billion. Congress ordered another redesign. In 1990, NASA engineers tried again. They shortened the dual keel, eliminated some equipment, and reduced the size of the laboratory and living mod-

President Ronald Reagan was a strong supporter of space station Freedom. *He predicted that it would enable scientists to make important advances in science and technology.*

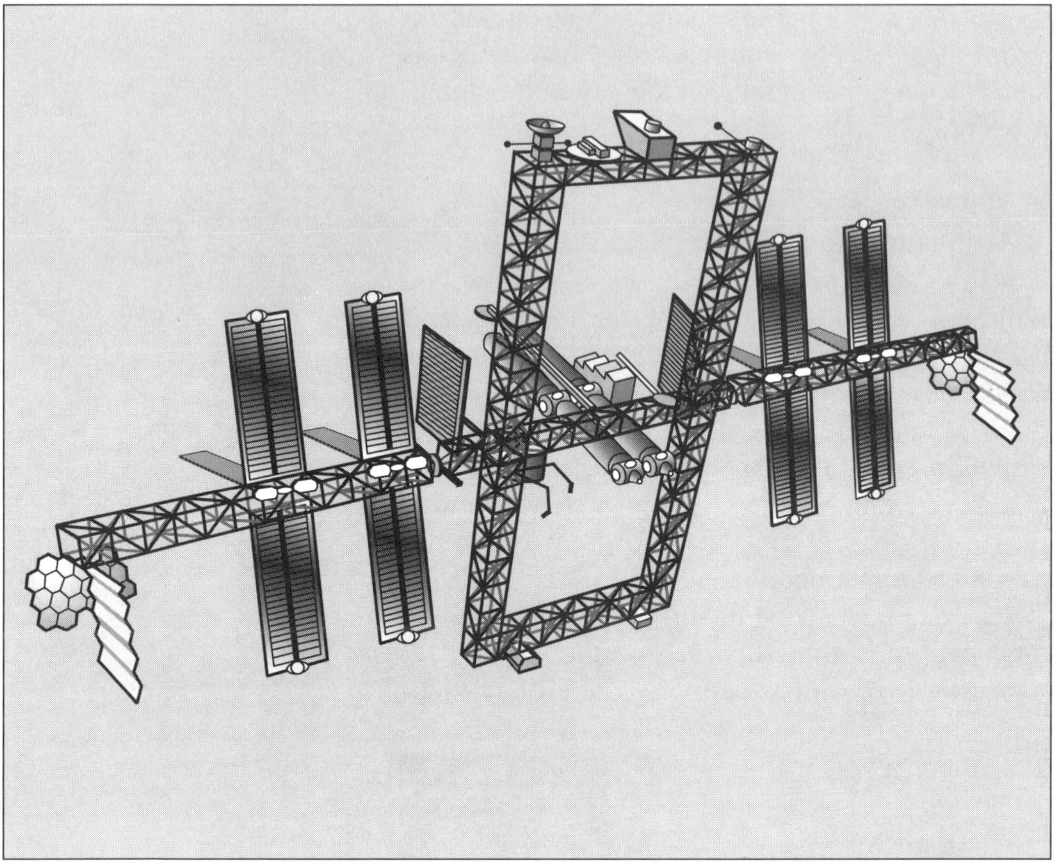

An illustration of NASA's original design of space station Freedom, *which was to measure more than 500 feet long. Rising costs have forced many design changes.*

ules. The new design cut the crew size to four and required more of the assembly work to be done on earth than in orbit. The estimated cost of the redesigned space station was $30 billion.

All these changes prompted much discussion about what *Freedom* should be able to accomplish. Some scientists feared that space station designers were trying to satisfy too many needs with the limited amount of money available for *Freedom.* They felt that a space station should be built, but that trying to make it serve too many purposes will not be in anyone's best interest. Apollo astronaut Michael Collins said:

> *Freedom* will be a multipurpose facility, and therefore a compromise. Some who intend to use

> the station as a laboratory want it designed for their particular experiments. Others want their own platforms in the sky and see *Freedom* as a competitor that will siphon off NASA's funds and attention, leaving them stuck on the ground. Without a unifying goal beyond the station, and without a compelling raison d'être [reason for being], *Freedom* has become an albatross around NASA's neck rather than a talisman of future success.

Is *Freedom* worthwhile?

Some scientists feared that the space station was being cut back so much that it would not accomplish enough to make it worth building. Many scientists declared that space shuttle missions and uncrewed spacecraft could provide them with much of the same data as the scaled-down *Freedom*—at far less cost.

The American Geophysical Union, a 27,000-member scientific organization, does not think that building a scaled-down *Freedom* is worthwhile. The organization contends that *Freedom*'s planned orbital path will only provide a limited view of the earth. Geophysical Union members believe that their goals of studying the earth from space would be better met by uncrewed earth-orbiting satellites placed into orbits where much more of the planet could be observed. The American Geophysical Union said of *Freedom*:

> Such poor coverage of the Earth and the lack of necessity for [human] observations are just two of the reasons . . . the AGU cannot give justification for the space station *Freedom*. It started out to be something good for everybody. We haven't heard of a justification yet that comes close to the amount of bucks we estimate the station will cost.

International partners with NASA—Canada, Japan, and the European Space Agency (ESA)—have committed $8 billion to research and development of *Freedom*. They are concerned that ongoing redesign of the space station will lessen its scientific and commercial value. These partners have invested heavily in *Freedom* as part of their space program, and do not feel that a scaled-down version will give them what they were originally promised. Ian Pryke, head of ESA's American office, said that ESA will "make it very clear that they want to be involved" in any future design changes.

Just as some scientists have worried about the adequacy of the design, others maintain that the United States does not even need to build a space station at this time. Many of these scientists feel that the funds that are planned to go into building, launching, and maintaining *Freedom* could be better used for other research projects. Many feel that the station is overpriced, and that the vast majority of research projects planned for *Freedom* could be more economically accomplished on earth, on robotic platforms, or on the space shuttle. In May 1991, the Council of Scientific Society Presidents said, "Scientific justification is lacking for a permanently [crewed] space station in Earth orbit."

Space station *Mir*

While many scientists see a space station as offering unique advantages for space research, some believe that NASA is duplicating efforts already begun by the former Soviet Union. A growing number of scientists believe that it would be in the best interest of space research for NASA to save the money it would spend on *Freedom* and work instead with Russia on the *Mir* space station.

The space station *Mir* was launched by the Soviet Union in 1986, and has been in almost constant use ever since. The Soviets used *Mir* for astronomical studies, as well as for experiments in

other fields of scientific research, such as studying the effects of long-term weightlessness on humans and other living organisms. Soviet cosmonauts occupied *Mir* for over a year at a time, far surpassing NASA's record of keeping astronauts in the *Skylab* space station for eighty-four days at a time.

Cooperative projects

By doing cooperative research projects with Russia on *Mir* instead of building *Freedom*, NASA could save billions of dollars, which could then be applied to other projects. With the cold war over, the United States and Russia, as well as many of the independent nations of what used to be the Soviet Union, have already begun to work together as scientific partners. Russian, American, and European researchers now freely exchange a great deal of information. Agreements have already been signed for a cosmonaut to fly on an American space shuttle mission, and for an American astronaut to fly aboard *Mir* in 1995. Yuri Semenov, a director in Russia's space program, is enthused about the increased cooperation and sharing of information between the two former rivals. He said, "This is a broad program of cooperation that is envisaged with hope. I look forward to it. The important thing is not to lose the momentum."

A large number of space scientists believe that

An artist's depiction of space station Freedom *in its orbit around earth. The space station is a proposed stopping-off point for missions to the moon or Mars, both visible in this illustration.*

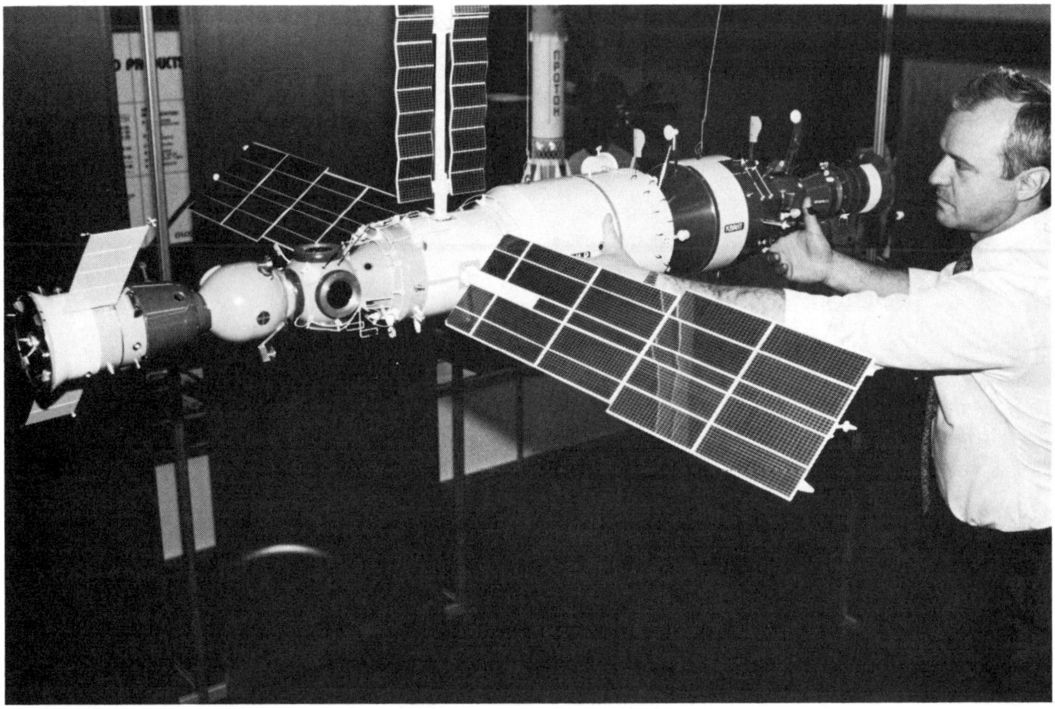

A model of space station Mir, *which is now being offered for sale by Russia.* Mir, *which has been occupied by astronauts for over a year at a time, could be bought by the United States for a fraction of the cost of building* Freedom.

most of the research projects NASA has planned for *Freedom* could be accomplished more economically by buying or leasing *Mir* from Russia, which has succeeded the Soviet Union as primary owner and operator of *Mir*. By doing this, the United States could have a ready-to-use space station at a much lower cost than that of *Freedom*. *Mir* is not all that American scientists would like for a space station to be, but NASA and the ESA could work with Russian scientists and engineers to make any needed modifications. Science writer Alex Roland believes that the United States would be wasting billions of dollars if it builds *Freedom* instead of purchasing the Soviet space station. He said, "The *Mir* space station is on the auctioneer's block; we could have a space station for a pittance if we choose."

There is also a strong contingent of scientists who believe that the United States must build a

space station if America is to continue its efforts to understand the universe. They say that if the United States is to continue to be seen as a world leader in technology, it needs to have a space station of its own. They feel that even though *Freedom* will not be as elaborate as many researchers would like it to be, it offers advantages beyond the space shuttles, space telescopes, and robotic probes now in use. Astronauts on space station missions would be able to perform experiments in life science, astronomy, and other branches of research that require longer periods of time to complete than NASA's current space shuttle missions allow. Axel Roth, deputy director of NASA, said, "There is no way I can envision that we will go forward in space without a space station. It's just a matter of whether we get it now or get it later. And it's going to cost a whole lot more later."

Mike Griffin, NASA associate administrator,

firmly believes that the space station *Freedom* can play an important role in preparing for future moon and Mars missions: "Space station *Freedom* will allow us to check things out in orbit before we have to rely on them for our lives. Wouldn't you rather check out your Mars habitat in Earth orbit before you actually get to Mars and find out you didn't get it quite right?"

Another new design

The tone and direction of the discussion concerning *Freedom* may soon change. Soon after President Bill Clinton took office in January 1993, he decided that the current design for *Freedom* was too costly, and would take too big a bite out of NASA's budget. White House science and technology advisor John Gibbons said, "When we looked at realistic numbers on space station *Freedom* spending, and the plausible limits of NASA spending a year or two down the road, we saw a wedge of station requirements that would have become a Pac-Man and eaten the rest of the agency."

President Bill Clinton directed NASA to again redesign Freedom. *The new space station is to be smaller, easier to build, and less expensive to operate.*

On March 11, 1993, NASA administrator Daniel S. Goldin announced that NASA was under orders from the White House to design a space station that was smaller, simpler, easier to build, and less expensive to operate than the most recent *Freedom* design. Goldin said that Clinton directed him "to redesign the space station as part of a program that is more efficient and effective and capable of producing good returns on our investment." In a speech to the American Astronomical Society, Goldin said that redesigning *Freedom* would mark the beginning of NASA's new focus on missions that are faster to prepare for and have quicker scientific payoff. He said, "If we continue with huge, long-term projects, the technology will always be outdated. Do we want to be up there with something in the year 2030 that was designed in the 1980s?"

In June 1993, President Bill Clinton announced that he would ask Congress to support a smaller, reduced-price version of *Freedom*. This version would cost an estimated $10.5 billion over the next five years. The cost of completing the new space station could reach $16.5 billion by the year 2001, which is still less than other recent designs. NASA has spent about $9 billion on the space station over the last decade.

Although Clinton asked NASA to shrink the size and cost of the space station, he stressed his support for a permanently staffed, orbiting laboratory. "We need to stay first in science and technology," Clinton said. "I think it would be a mistake to scrap all the work we've done."

Freedom's features

The new design, which NASA planned to complete by fall 1993, will use many of the features already developed for *Freedom*. These features include pressurized living and laboratory areas and solar power units. Living and lab areas and solar panels will be attached to a backbone-like structure that will extend about 300 feet. The space station will permanently house a crew of four astronauts.

The president's proposal followed four months of redesign work by NASA engineers and others. After completing its work, NASA presented Clinton with its three top choices for redesigning the space station. Clinton directed NASA to combine two of those choices to create a new design. NASA's international partners in Japan, Canada, and Europe will also help work out details of the redesigned space station.

Members of Congress who have opposed the space station before say they will continue their

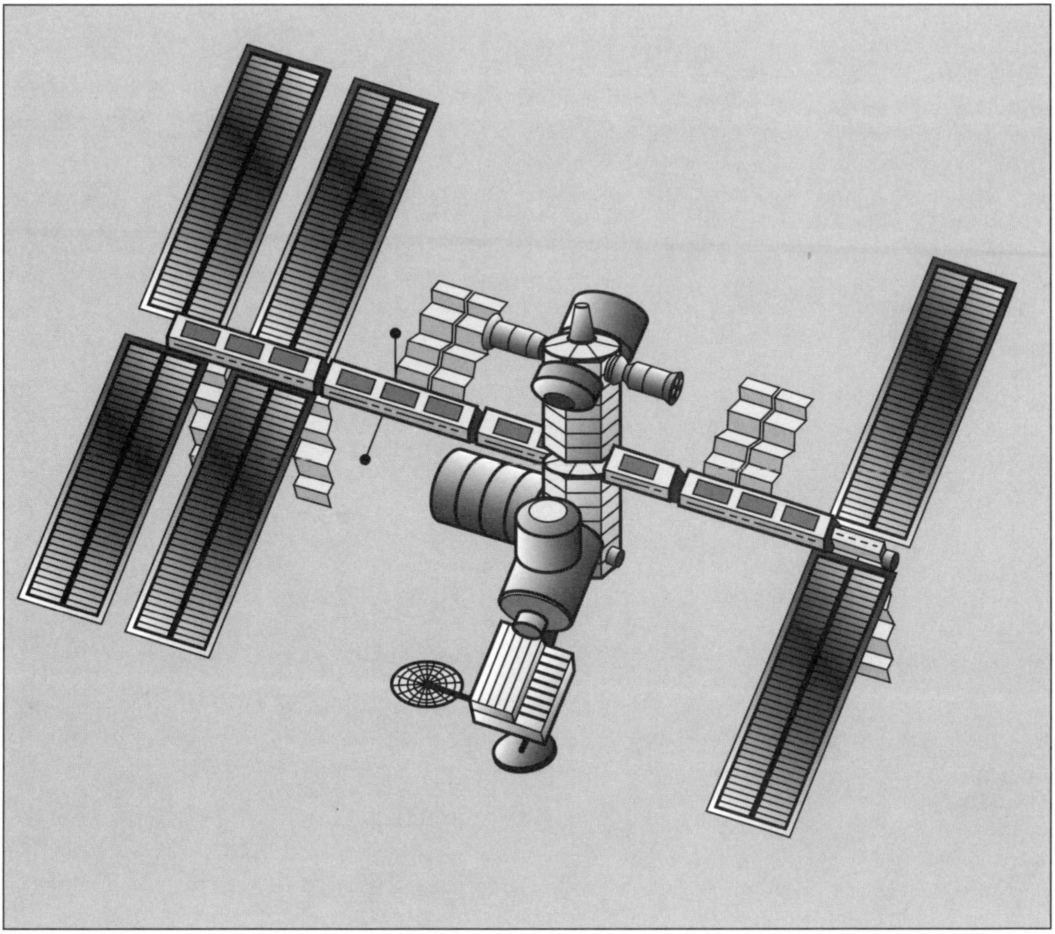

fight to kill the project. Some space station critics say they believe the project still costs too much for the results it will bring. Space station supporters are expected to rally behind it, however. Because views on this issue are so strong, debate probably will continue for some time.

An illustration of NASA's scaled-down version of space station Freedom. *Proponents of the space station still face a difficult fight against opponents who insist that the project is not worth the money.*

USA
NASA

5

Is a Moon Base Needed?

(Opposite page) An artist depicts astronauts exploring the moon's rugged terrain in a lunar rover. While a moon base would make such exploration possible, many debate whether such a base is necessary.

DURING THE 1960s, NASA's main goal was to send humans to the moon—and to do so before the Soviets could. It was a goal backed by many of the country's top scientists, rocket engineers, and government officials. Apollo astronaut Edwin E. Aldrin Jr. said:

> The greatest technical achievement of Apollo probably was the development of practical interplanetary exploration organization that combined the resources of NASA, the commercial sector, and our university laboratories. America answered John Kennedy's challenge and learned how to explore the solar system.

But in the early 1970s, only a few years after the first humans had set foot on the moon, the mounting expense of America's military involvement in the Vietnam War forced Congress to drastically cut many federally funded programs—and NASA's budget was one of the hardest hit. Once the last crewed lunar mission, *Apollo 17*, was completed in 1972, NASA's lunar exploration came to a virtual standstill. Isaac Asimov, the famous science writer, said, "After *Apollo 17*, it became clear that the attitude toward expeditions on the moon was, 'We scored a touchdown.

Astronaut James B. Irwin, behind the wheel of a Lunar Roving Vehicle during the Apollo 15 *mission in 1971. Since the final Apollo mission in 1972, human exploration of the moon has been at a standstill.*

We won the game; now we can go home.'"

Budget constraints halted plans for pushing ahead with a moon base. Ambitions for expanding human presence in space were cut back to the modest goal of launching and crewing the small earth-orbiting space station, *Skylab*. Former Cable Network News (CNN) bureau chief William J. Walter said, "The Apollo program was originally intended as a starting point for the human exploration of space, but instead it became a finishing line . . . after the first flurry of missions, nothing. No base, no observatory, not a single piece of hardware to point the way to the future."

Why return to the moon?

Much remains to be learned about the moon. One group of scientists would like to study the moon as a way of learning more about the history of the earth—and the history of the solar system.

Because the moon has no atmosphere or running water, its geology has remained essentially undisturbed for billions of years. Astronomer Robert Jastrow believed the moon is a living laboratory, and said, "The moon offers a chance of recapturing the vanished record of those first billion years of the history of the solar system. It can help us solve the mysteries of the origin of Earth and the origin of life." Scientists would like to use the moon as an observatory and a science lab. Many would also like to see a moon base constructed as a nearby practice ground for future human flights to Mars.

Scientists from NASA can gather some of the scientific information they seek with uncrewed probes. The first lunar spacecraft to be launched by NASA in over twenty-five years are now be-

Apollo 10 *provides a view of the moon's surface and the earth rising on the horizon. Many scientists believe that study of the moon can teach them much about the history of the earth.*

ing developed. Two uncrewed Lunar Resource Mapper probes are scheduled for launch in 1995 or 1996. The probes are expected to relay data telling scientists the locations of the best deposits of minerals and other substances. The cost for the project is expected to be about $100 million. One purpose of the Lunar Resource Mapper is to demonstrate NASA's new management technique to Congress. As NASA Exploration Office manager Humboldt C. Mandell Jr. said, "It is to demonstrate that NASA can accomplish things less expensively and faster than we ever have in the past."

To continue study of the moon, NASA also hopes to launch a $100 million uncrewed Common Lunar Lander in the late 1990s. It plans to ask Congress for $31 million to initiate the program. These uncrewed lunar missions are NASA's first exploratory steps toward sending humans back to the moon.

A permanent moon base, where humans could live for weeks—even months—and carry out long-term experiments is in NASA's long-range plans for lunar exploration. The goal is to have at least a small moon base functioning by the year 2000, and have it permanently occupied by 2010. The cost of building a small moon base, spread over two decades, would be about $80 billion. Mission planners at the Johnson Space Center expect that the first crew of astronauts at the base, which will be called the First Lunar Outpost, will be able to stay on the moon for about six weeks at a time. As the base develops, later crews will be able to remain on the moon for stays of several months or more.

Limiting expenses

To make the project appealing to Congress, NASA plans to do everything possible to limit expenses. The equipment used at the First Lunar Outpost will borrow from hardware that has already been developed for other NASA missions. Some of the hardware designed over twenty years ago for the Apollo moon missions will probably be incorporated into the design.

Renowned scientist and writer Arthur C. Clarke believes that a moon base would be a worthwhile scientific investment:

> The moon will also be of the utmost importance, not only as an object of study, but also as a base for an enormous variety of experiments. Everything that has been said about the value of space stations for physical, biological, and astronomical research also applies to the moon.

An artist's vision of what a moon base might look like. Such a base would allow astronauts to conduct long-term experiments on the moon.

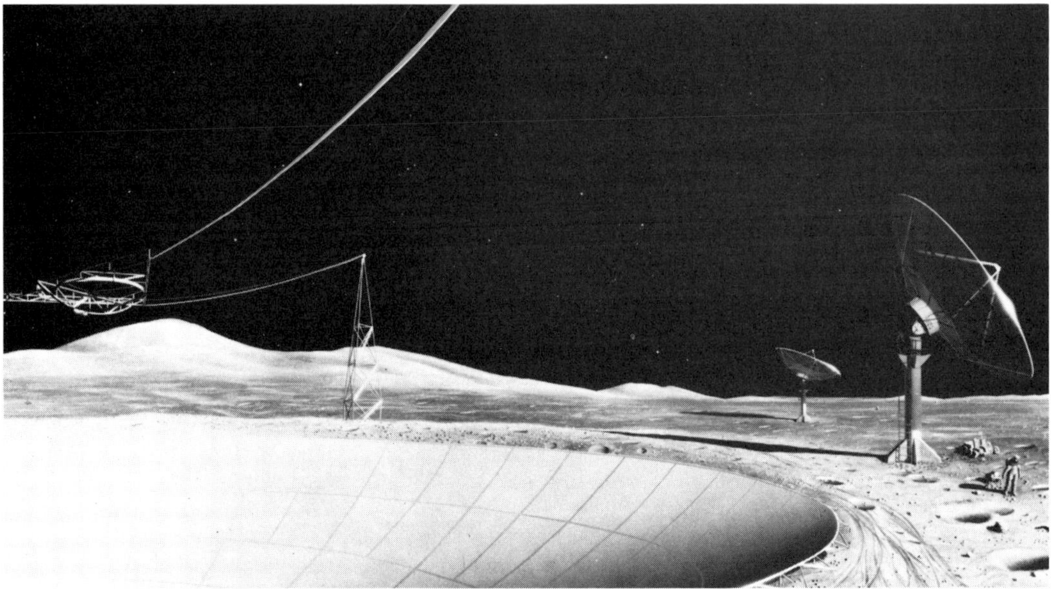

Optical telescopes on the moon would not be hampered by lights and atmospheric distortion that get in the way of earth-based telescopes.

Some scientists push for a moon base because of the opportunities it would present in lunar geology, and others want a moon base because it is an ideal location for both optical and radio telescopes. Optical telescopes on the moon would not have to overcome atmospheric distortion or light pollution. The fact that the moon's gravity is only one-sixth that of earth means larger telescopes with heavier lenses can be used without collapsing under their own weight. Optical telescopes with an aperture of about 82 feet have already been proposed. They would have 100 times the light-gathering power of the Hubble Space Telescope and would provide a direct view of planets orbiting distant stars.

Scientists working with lunar radio telescopes would no longer have to sift through the endless stream of radio signals generated by humans to find the information they want. According to Arthur C. Clarke, "The very center of the far side of the moon is the only place in the whole solar system that is permanently shielded from our noisy planet—by 2,000 miles of solid rock."

Another important reason astronomers would like to see telescopes built on the moon is that the moon rotates at a much slower rate than the earth. The moon turns on its axis only once every 27.3 days. This slow rate of rotation will give astronomers of the future the advantage of being able to view celestial objects continuously for up to two weeks at a time. Because the moon has no atmosphere to scatter light waves, astronomers on a moon base will even be able to use light-gathering telescopes in the daytime—as long as the objects they want to study are not too close to the sun.

Using lunar materials

Scientists also want to use a moon base as a laboratory to find ways of helping future space travelers make the best use of available extraterrestrial materials. The Apollo astronauts brought back over 1,000 pounds of moon rocks and soil in over two thousand separate samples taken from various locations. Moon rocks are known to contain rare minerals and other substances that could be useful for improving life on earth. Helium 3 (He-3), which is rare on earth, was found to be plentiful in samples of lunar soil. It can be used as a fuel in nuclear fusion reactors, which produce energy by joining atoms. Former Apollo astronaut Charles D. Walker said, "The moon may be the site for truly cheap, inexhaustible energy. Lunar helium-3 could provide us with an environmentally acceptable and possibly inexhaustible source of fusion energy."

Scientists estimate that at least a million tons of He-3 are readily recoverable on the lunar surface. "Buzz" Aldrin believes that the possibility of using lunar He-3 to meet the mounting need for clean-burning fuel here on earth should be fully explored. Aldrin states, "Nuclear fusion re-

A lunar rock sample brought back by the crew of Apollo 17. *Researchers hope to make use of moon rocks and other lunar materials for building a moon base.*

actors powered by He-3 would be almost completely devoid of radioactivity. And much of the Earth's future energy needs could be met by burning only 200 tons of He-3 a year in environmentally benign fusion reactors."

On a moon base, astronauts could experiment with lunar soil and rocks to find ways of extracting minerals, liquid, gases, and other useful substances. They would learn how to use lunar materials to build shelters and other structures so that construction materials would not have to be ferried from earth to the moon or from earth to future colonies on other planets.

Eventually, lunar mining could supply the raw materials necessary for manufacturing a wide variety of products. Moon rocks are known to contain materials that can be refined to manufacture aluminum, magnesium, silica, chromium, and iron. Oxygen, which is trapped in lunar rocks and soil, could be extracted and used for air and fuel. It can also be combined with hydrogen, which would be transported from earth, to create water. Because of the moon's low gravity, tanks of moon base-extracted oxygen could eventually be rocketed to space stations much more economically than tanks launched from earth. Scientists speculate that craters near the poles of the moon contain small amounts of frozen water—possibly frozen for billions of years. If true, this supply of lunar water could be defrosted and used to meet the needs of astronauts stationed at the moon base.

An artist depicts a robotic unit extracting gases from lunar materials. Scientists predict that lunar mining could provide astronauts with air, water, and fuel.

A practice ground

A moon base would also be used as a practice ground for future crewed missions to Mars. Tech-

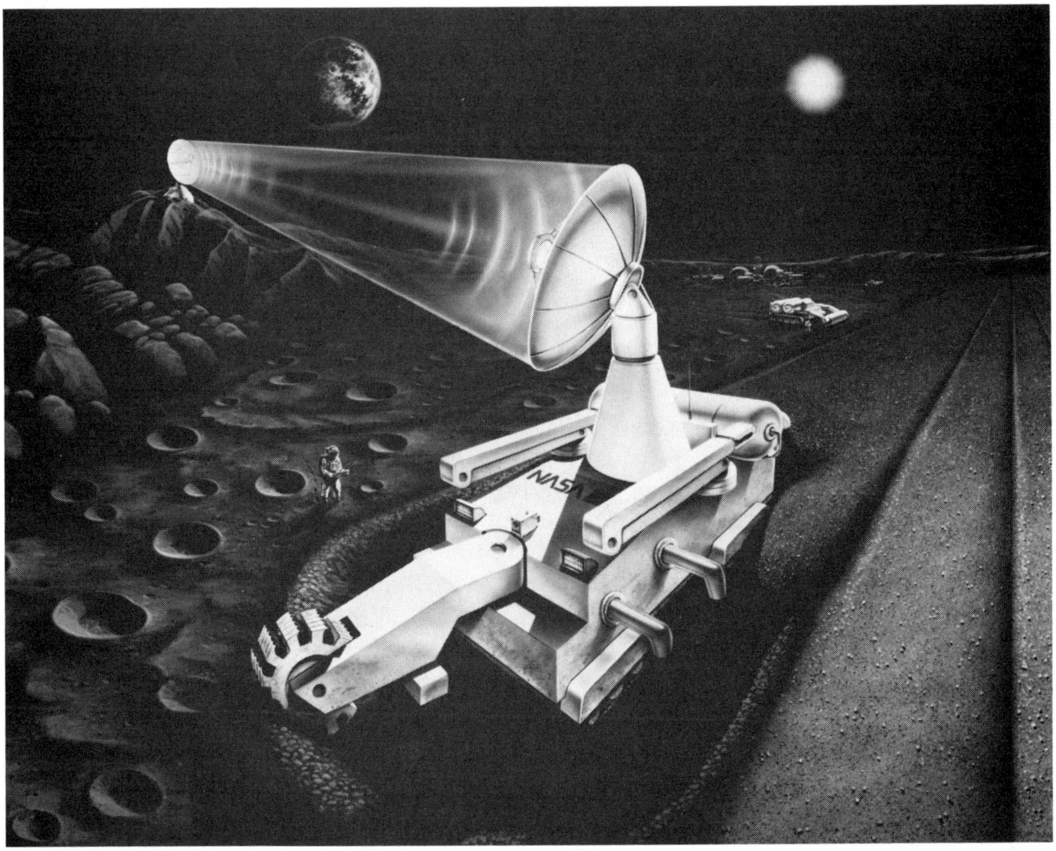

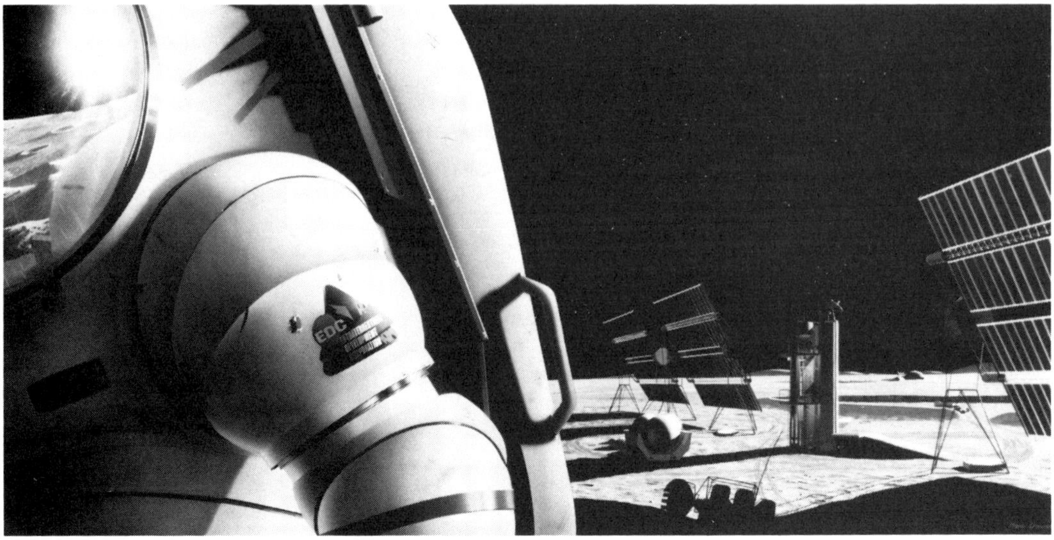

Only a three-day trip from earth, a moon base could be used as a practice ground for future human flights to Mars.

nology could be tested in a low-gravity environment that is relatively close to earth. The moon is only a three-day trip from earth, but a return trip from Mars would take space travelers nearly a year. Thus, astronauts on a moon base would have a chance to solve any equipment problems they encounter before they are too far away from home.

A 1987 NASA task force headed by astronaut Sally Ride recommended in a report that NASA build a moon base before trying to send humans to Mars. The report stated:

> Colonizing Mars should be our eventual goal, but it should not be our next goal. Sending people to and from Mars is not the only issue involved. Understanding the requirements and implications of building and sustaining a permanent space base on another world is equally important. We adopt a strategy of natural progression which leads step by step, in an orderly, unhurried way, inexorably toward Mars.

Opinions are not unanimous on the need for a moon base as a first step toward the exploration of Mars. Apollo astronaut Michael Collins believes that NASA can do its preliminary research more quickly and at less cost on space station *Freedom* or *Mir*. Collins said, "I reject the whole Moon [base] idea. I think it is an unnecessary detour on the way to Mars, and one that could gobble up the entire Mars budget." Carl Sagan agrees. Sagan said, "A lunar base wouldn't be a detour on the road to Mars, but a trap. We would use up financial resources and indefinitely delay going to Mars. Mars is so much more exciting."

Improving launches

When NASA is ready to send crewed missions to the moon to set up a lunar colony, it will need more powerful rockets than are presently available. The huge Saturn V rockets that NASA used to send Apollo astronauts to the moon during the

Astronaut Sally Ride and others believe that a moon base is a necessary first step toward exploration of Mars.

1960s and 1970s are no longer in production. The rockets currently available to NASA now are much smaller than the Saturn V. They are not capable of lifting the air, water, shelters, and other supplies necessary for extended lunar missions. Because of this, NASA officials are looking at the possibility of leasing Russian Energina heavy-lift boosters. The Energina rockets have an 8-million-pound thrust and can send payloads weighing up to 70,000 pounds to the moon.

The National Launch System (NLS), currently being designed by NASA and the Department of Defense, will be the first significant improvement

Before the United States can build a lunar colony, NASA will need more powerful rockets than those currently being used.

in American launch capabilities since the 1960s. The project, if approved by Congress, is estimated to have a total price tag of $11.5 billion. Robert Cooper, president of Atlantic Aerospace Electronics, said, "The NLS is something that has been delayed for many, many years. As a country, we have needed to refurbish our launch capacity." Most of the rockets that are currently being used for uncrewed payload launches employ missile technology that is twenty-five to forty years old. "We have been supporting our whole space program with modifications to old ballistic missiles, and that has been enormously expensive," Cooper said.

Although most NASA officials agree that a new launch system should be built, there is some controversy over which purposes it should be designed for. Some want the new system to serve primarily as a satellite and payload launcher. Others hope that the NLS focuses more on large heavy-lift rockets that can put payloads weighing up to 80,000 pounds into orbit. Such a rocket could be used to resupply a space station with water, food, and equipment. Still other future-thinking scientists would like to see the NLS include rockets that are capable of boosting 200 metric tons of payload into space. These rockets would have enough power to launch the massive components needed for a moon base or a crewed Mars mission—and serve NASA's needs well into the next century.

6

The Goal of Exploring Mars

(Opposite page) An artist's vision of a research station depicts vehicles, living areas, greenhouses, a launching and landing area, water pumping station, and maintenance garage. These and other structures would be needed by any research base on Mars.

BECAUSE MARS IS the most earth-like planet in the solar system, it holds a special fascination for humankind. At one time, scientists thought they had found evidence on Mars of a breathable atmosphere, an abundance of water, and the ability to support some form of life. Uncrewed probes sent to Mars during the 1960s proved those theories to be incorrect. Mars is a dry, arid planet with an atmosphere only about one-hundredth as dense as that of earth. Such an environment cannot support life as we know it.

However, photographs taken by NASA probes show evidence of a different environment on Mars millions of years ago. Dry riverbeds, lakebeds, and oceans could be clearly identified on the martian surface. At one time in its ancient past, scientists speculate, Mars might have had an Earth-like atmosphere and could have even supported some primitive forms of life. Astronomer Carl Sagan said, "*Mariner 9* and *Viking* found hundreds of dry riverbeds, apparently indicating a time in the recent geological history of the planet when abundant liquid flowed. It is a world awaiting exploration."

To continue exploration of Mars, NASA

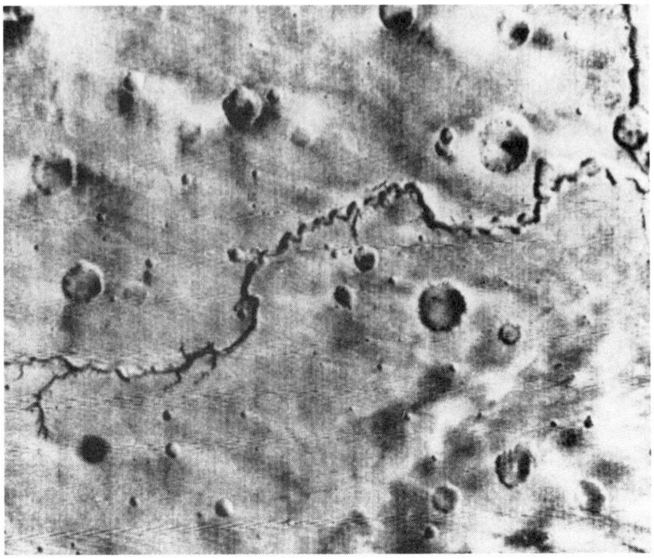

This photograph of a meandering "river," taken by Mariner 9, *is the most convincing evidence that liquid once flowed along the surface of Mars.*

launched the *Mars Observer*, an uncrewed probe, in September 1992. The $900 million project was NASA's first mission to Mars in seventeen years. The *Mars Observer* was to travel through space for eleven months in order to reach its orbit around the red planet by late summer 1993.

Relaying information from Mars

During its 687-day journey orbiting Mars, the *Observer* will constantly relay information about Mars back to earth. The planet's weather systems, surface composition and landscape, and magnetic and gravitational fields will be studied, as will the supply of frozen water at Mars's polar regions. The *Mars Observer* will study how natural forces interact on Mars to cause environmental changes. It will also provide the preliminary data necessary for scientists to select the best landing sites for future Mars missions.

The *Mars Observer* will gather and relay more data about Mars than has been possible in NASA's past missions. "As with the Earth Observing System planned for the late 1990s, the

Mars Observer will give scientists a systematic body of data gathered about the same planet," said Arden Albee, project scientist. "While no single instrument will make an incredible discovery, all the instruments will work together to allow us to test our model of how Mars works."

One of the *Mars Observer*'s assignments is to take a closer look at the controversial "face monument" on Mars. Some scientists speculate that this rock formation, which was first photographed in 1976 by NASA's *Viking* space probe, is some kind of a stone monument built by a long-gone civilization. Richard C. Hoagland, an astronomer, news correspondent, and writer who has worked with numerous space-related projects, has been one of the major advocates of NASA's taking a closer look at the monument. Whereas some scientists believe that it could indeed be a monument created by civilized beings, other experts who have studied the pictures conclude that

The rock formation (far right) photographed by the Viking *space probe, resembles a human face. While probably only an optical illusion, some people speculate that it is a stone monument built by an ancient civilization.*

it is only an optical illusion that appears to resemble a face when sunlight strikes it at the proper angle. Since the *Mars Observer* is equipped with cameras that have resolution five times better than that of the *Viking*'s cameras, it will give astronomers clearer pictures on which to base their opinions.

Officials hope to launch a ten-year Mars Environmental Survey program, sending a new probe, *Pathfinder*, into orbit in 1996. It will gather even more detailed information about the planet's geology and environment. Beyond that, NASA's future exploration of Mars is uncertain. Many NASA scientists believe that the next step should be launching return-sample missions that would scoop up martian soil from different parts of the planet and return it to earth for study.

Sending humans or robots

One of NASA's long-range goals is to send humans on missions to Mars. But before humans can be sent on Mars missions, many technical and environmental problems must be solved. To successfully complete a human Mars mission, artificial environments must be developed that would supply astronauts with sufficient oxygen, temperature control, and radiation protection. A crew of eight astronauts would need as much as 75,000 pounds of oxygen, food, and water to keep them alive during a two-and-a-half-year Mars mission. A rocket large enough to carry the astronauts, supplies, and fuel to Mars could easily weigh as much as 1 million pounds. Even if supplies for the mission were lifted into earth's orbit in sections, heavy-launch rockets, which NASA does not currently have, would be needed to accomplish the mission.

Because of the distance and travel time involved, some scientists believe that human flights

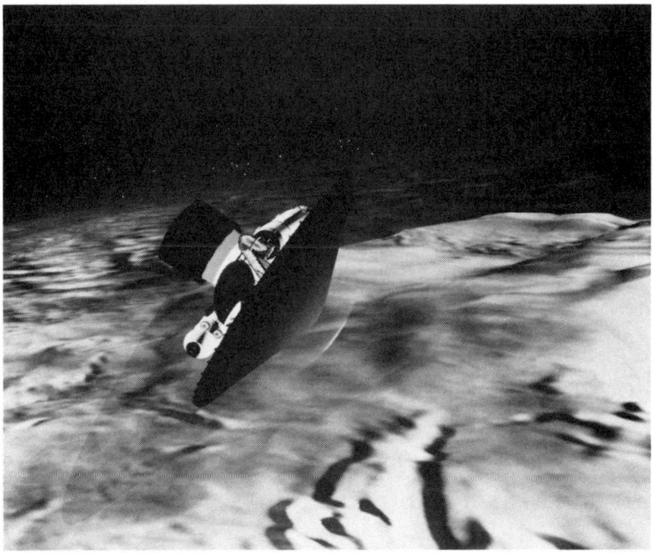

In this illustration, a piloted space vehicle performs a maneuver over the martian surface. By the time missions to Mars are possible, some people say that robots will be able to perform human tasks at less cost and with less risk.

to Mars are unnecessarily dangerous and expensive. They contend that by the time NASA is ready to send astronauts to Mars, scientists will have developed robots that are capable of accomplishing almost any task humans can at much less cost and without the risk of life. Science writer Alex Roland said, "Any specific function that can be identified in space can be done more cheaply and more reliably by machines than by people."

But there are drawbacks to a Mars mission that requires robots, because they would have to be given instructions from earth. Depending on the position of the planets, radio signals take anywhere from five to twenty minutes to travel from Mars to earth—and that long again to travel back from earth to Mars. The time delay between when a command is given and when it is received would make it very difficult for scientists to control a Mars-based robot, so robots used for such missions would have to have some level of artificial intelligence. "With a dumb robot, by the time the operator on Earth realizes that a machine was heading off a cliff, it would be too late to save it," said James D. Burke, technical editor of the *Planetary Report*. "That is why Martian robots need to be smart."

Even if robots could be built that are able to come close to doing the work of a human, many scientists feel that the goal of sending astronauts to Mars—and thus expanding human presence in the solar system—would be worth the extra planning, expense, and risk involved. They believe that sending humans to Mars is a goal that would not only increase scientific knowledge, but also hold public interest—and therefore congressional interest in funding preliminary projects in this century that would make a human Mars mission possible in the next century. Apollo astronaut Michael Collins said, "A Mars project would rejuvenate NASA, and would carry in its wake a new generation of scientists and engineers with the education necessary to succeed not only in space careers, but in other high-tech endeavors as well."

An artist depicts a robot repairing equipment on the Freedom *space station. Even if robots can do the work of humans, many people believe a human presence in space is worth the extra risk and expense.*

Convinced that human exploration of Mars would reveal a scientific gold mine, many researchers have urged NASA to make human flights to Mars a high priority.

The Planetary Society, headed by Carl Sagan, is one of the most vigorous proponents of sending humans to Mars. For several years now, the 125,000-member public group has been a leader in pushing for the government to make human flights to Mars one of NASA's priorities. To show widespread support for the goal, they circulated "The Mars Declaration," a document that set forth reasons for pursuing human exploration of Mars. It has now been signed by over 150,000 people. Two hundred and fifty prominent Americans have signed the document—including former president Jimmy Carter; former United Nations ambassador Jeane Kirkpatrick; news anchor Walter Cronkite; and the astronauts of *Apollo 11*, who took part in NASA's first human moon landing.

Human exploration of Mars

"The Mars Declaration" proclaims that human exploration of Mars would reveal a scientific gold mine, making it well worth the expense. Scientists could study climatic changes on Mars and search for past or present life. Application of

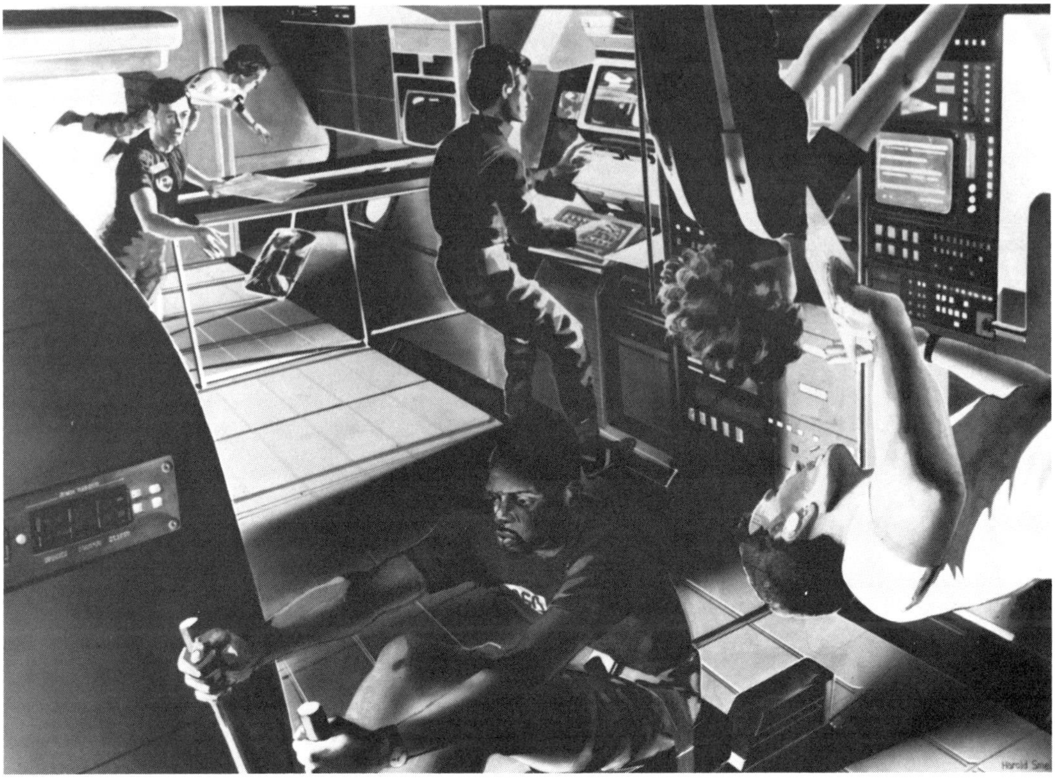

Space station Freedom *could be used as a study base for gaining knowledge necessary for future human exploration of Mars. The effects of long-term weightlessness on humans is one area that requires more study.*

whatever knowledge was gained from the missions could be used to better understand the earth and its environmental systems. The Planetary Society also views a crewed Mars mission as a way of reviving the stagnant U.S. planetary space program, and providing a focus and sense of purpose for NASA. It believes that having a crewed Mars mission as a long-term goal would give NASA a solid reason for building the space station *Freedom. Freedom* would be used for in-orbit assembly of space vehicles for studying long-duration life support systems and the effect that long-term weightlessness has on humans.

Apollo astronaut Michael Collins said, "There is a scientific bonanza waiting on Mars: comparative planetology, the scientists call it. They would like to study the history of the Solar System and the evolution of the planets by comparing evidence found on the Earth, the Moon and Mars." By studying ice core samples, for example, scientists can learn many things about a planet's past. Ice samples taken from the martian poles could tell the story of the planet's climate changes, just as samples of ice from the earth's arctic and antarctic regions tell the story of earth's climatic history.

Mars colonies

Early Mars colonies would be small. They would be made up of several astronauts living in modest shelters. The first human visitors to Mars would probably stay there for less than six months. They would depend mostly on supplies rocketed to Mars from earth. On later missions, as scientists gain experience in processing and using martian resources, the colonists would be able to spend more time on the planet as they depend less and less on materials transported from earth. The atmosphere on Mars is very thin. It

A Viking *photograph of the rocky surface of Mars. Scientists might learn more about the history of the solar system by studying ice samples from the martian poles.*

does contain oxygen and all of the other gases necessary for maintaining life, but not in the right proportion or density to meet the needs of human beings. Equipment can be brought to Mars that can take in the thin martian air, adjust the balance of gases, and pump the new Earth-like air into the colonists' pressurized shelters.

Sharing the cost

Sending humans to Mars—even for a short stay—would be a tremendously expensive venture. To offset costs, the human flight to Mars would probably be a combined world mission, financed jointly by the United States, Russia, the European Space Agency (ESA), and a number of other countries. Because the round trip would take between two and three years, huge quantities of air, water, and food will have to be transported to keep the astronauts alive on the trip. Artificial gravity for the spacecraft might also have to be designed.

To send humans to Mars, even by the most conservative estimates, would cost at least ten times more than sending an uncrewed craft. Uncrewed flights to Mars cost hundreds of *millions* of dollars, and crewed flights are estimated to cost hundreds of *billions* of dollars. At the present time, NASA estimates that a crewed Mars mission would cost in the neighborhood of $400 billion—and approximately $25.7 billion of that would just be for launching the 2.2-million pounds of fuel needed.

In spite of all the obstacles, many NASA officials believe that this long-range goal is necessary to give the space program direction and to help push it ahead into the next century. Michael Collins said, "NASA needs a new place, Mars—and needs it badly. It's not just the money NASA needs, it's having a goal, a unifying vision for the

One artist's vision of a moon base. Not all scientists favor construction of a moon base. Some believe that money and time spent on a moon base would be better spent going directly to Mars.

future. A healthy, vigorous space agency is a national treasure that deserves to be nurtured."

Picking a plan

Scientists who are in favor of a human Mars mission often disagree on how best to accomplish it. At the present time, a number of plans are being considered.

The plan most favored by NASA officials involves first gaining more experience with space stations, then building a moon base before attempting to send humans on to Mars. This plan has the advantage of giving scientists a chance to test many kinds of technology in near-Earth situations, so astronauts would have a better chance of survival if something malfunctioned. Astronauts aboard the space station would be only a few hours from earth, and astronauts on the moon would be only three or four days away—but it would take astronauts on Mars nearly a year to return to their home planet. John M. Logsdon, a NASA advisor, said, "Establishing a permanent base on the moon is an essential stage of expansion beyond Earth. Humans should go to Mars, but it does not make sense to bypass the moon, a celestial island just offshore that offers significant knowledge for colonizing space."

The disadvantages of the "moon first" plan are the time and tremendous expense that would be required to complete both a space station and a moon base before even starting to work seriously on a crewed Mars project. That is why some scientists feel that after some experience with an American or Russian space station, NASA should make plans to send humans directly to Mars without going to the expense of building a moon base. They say that a moon base would take too long to build and consume too much NASA money, thereby slowing down human exploration of Mars and other space research projects. Although

A fully enclosed Mars research base might look something like this.

there would be more risk in sending humans to Mars without moon-base preparation, those in favor of the plan contend that, as in all adventures into the unknown, there will always be an element of risk.

Some scientists go even further in their belief that humans can realistically go to Mars in the near future—possibly before the end of the century. They believe that the technology can be developed here on earth to send astronauts directly to Mars, skipping both the space-station and moon-base stages. One such plan, "Mars Direct," was proposed by a team of scientists at Martin Marietta Astronautics in Denver, Colorado. Their plan would cost about $50 billion to put into action as opposed to NASA's current $400 billion plan. The "Mars Direct" approach involves first launching a supply-and-shelter ship and placing it in orbit around Mars. Once it has safely arrived, another rocket carrying the astronauts would be launched. A combination of resources transported from earth and resources found on Mars would be used, which would help make the "Mars Direct" mission much more cost-effective than the entirely Earth-supplied mission NASA plans.

First, faster rockets

The length of time it takes to make a trip to Mars and back is one of the major difficulties NASA scientists and engineers encounter in planning human flights to the red planet. This is why many space experts believe that designing and building better propulsion systems should be a high priority of the space program. If faster engines are developed and if travel time can be shortened, astronauts will not have to endure zero-gravity conditions for such long periods. Shorter journeys would reduce storage and heavy-launch problems, since less water, air, and food would have to be taken on the journey. Danger to the astronauts from space radiation would also be reduced by reducing the exposure time.

Colonizing other planets has always been a long-term goal of NASA's crewed space program. If astronauts are to go beyond Mars, perhaps to the moons of Jupiter and Saturn, they must be able to travel much faster than they can with rocket engines that are presently available. Scientists and engineers at NASA, as well as in the private sector, are working to develop better and faster means of propulsion.

One of the most practical ideas for improved rockets is nuclear engines. Nuclear engines have the advantage of requiring much less fuel mass than chemical rockets. This makes them lighter and easier to transport into space. They can be fired for much longer periods, and give a spacecraft much more thrust and speed. Since nuclear rockets would only be operated when the spacecraft is beyond low-earth orbit, many scientists feel that they would not pose a threat to humans or the environment. However, some scientists worry that if a rocket malfunctioned or exploded, as in the *Challenger* space shuttle tragedy, dangerous amounts of radiation could be released into the atmosphere.

Possibilities for the future

Even with nuclear-powered rockets, interplanetary trips would be time-consuming and expensive. Some space scientists, therefore, are already looking beyond nuclear energy to the possibility of antimatter propulsion. Antimatter is made up of atoms with the exact opposite electrical charge of ordinary matter. When antimatter particles and matter particles are combined, they destroy each other, leaving behind intense bursts of energy. Science writer Gregg Easterbrook said:

> Preliminary speculation is that mid-twenty-first century antimatter technology may enable the construction of machines moving at about 1 percent of the speed of light—nowhere near fast enough for exploration of the closest star . . . but plenty fast enough for operations among this solar system's planets.

Humans may one day have the technology to travel at the speed of light, making space travel and exploration possible to a degree never before imagined.

Astronomer Robert Jastrow also believes that antimatter technology can answer the needs of future space programs:

> Antimatter used as a rocket fuel is hundreds of times more powerful, pound for pound, than the fuel used in nuclear fusion. If copious quantities of antimatter could be manufactured on the earth and fed into an engine at a steady, controlled rate, we would have a rocket capable of approaching the speed of light.

Glossary

astronomical research: Space research.

big bang theory: The theory that the universe began with a large explosion over 15 billion years ago.

electromagnetic radiation: Waves of constantly changing magnetic and electric fields that produce light waves, sound waves, X rays, gamma rays, and others.

European Space Agency (ESA): A space agency run jointly by Austria, Belgium, Denmark, France, Germany, Holland, Italy, Norway, Spain, Sweden, Switzerland, and the United Kingdom.

frequency: The number of electromagnetic waves, such as light and sound waves, that can pass a given point per second.

infrared radiation: Radiation that has a wavelength slightly longer than that of visible light.

life science research: Research conducted on plants, animals, and microorganisms.

light-year: The distance light can travel in one year, about 5,900 billion miles.

materials research: Research conducted on crystals, metals, and other nonliving substances.

microgravity: A near-zero gravity level experienced by astronauts once they are far enough away from earth to be free of most of the planet's gravitational pull.

microwave radiation: High-frequency radiation that lies between radio waves and infrared waves on the frequency spectrum.

Mir: The Russian space station.

moon base: A facility where humans can live on the moon for extended periods of time.

National Aeronautics and Space Administration (NASA): The federal agency that oversees and conducts U.S. space exploration and research programs.

National Launch System (NLS): A new launch system NASA and the Department of Defense are designing for better launches throughout the space program.

nuclear rockets: Rockets powered by any form of nuclear energy.

optical telescope: A telescope that magnifies light waves.

payload: The load carried by a spacecraft; consists of astronauts, scientific instruments, and other items that relate directly to the purpose of the flight.

pulsar: A star that emits short, precisely timed bursts of radio signals.

quasar: A powerful source of radio energy that is speculated to be in the center of a distant galaxy.

radio telescope antenna: An antenna that receives and amplifies radio waves from space.

sample-return mission: A mission that involves collecting a soil sample from a planet and returning it to earth.

satellite: An artificial object put into orbit around the earth.

Saturn V rocket: Rocket used to launch the Apollo moon missions of the 1960s.

shuttle: A reusable spacecraft in which astronauts conduct experiments and observe astronomical phenomena; also used for launching satellites or experimental space-based weapons.

Skylab: NASA space station used in the mid-1970s.

smart robots: Robots that are programmed to solve problems without constant human input.

solid rocket booster: A rocket that uses explosive powder for fuel. It is usually used to help lift a spacecraft off the ground.

space colony: A facility constructed away from earth that has the environment necessary for human survival.

space race: The competition between the United States and the Soviet Union to excel in space during the cold war years.

space station: An earth-orbiting facility where humans can live and conduct research for extended periods of time.

thrust: The power a rocket has to pull away from earth's gravity.

ultraviolet radiation: Radiation at frequencies between those of X rays and visible light.

Voyager: Two crewed interplanetary probes launched in the 1970s to take close-up pictures of the outer planets in the solar system.

X rays: Radiation at frequencies between those of gamma rays and ultraviolet rays.

Organizations to Contact

The following organizations are involved with issues concerning space exploration. Most offer magazines or brochures on the topic.

American Institute of Aeronautics and Astronautics (AIAA)
370 L'Enfant Promenade SW
Washington, DC 20024
(202) 646-7400

AIAA is made up of aeronautical and astronautical scientists and engineers. Many of its members support crewed space exploration, the space station *Freedom*, and crewed missions to Mars. The AIAA publishes the monthly magazine *Aerospace America*, which covers new developments in space technology.

National Aeronautics and Space Administration (NASA)
600 Independence Ave. SW
Washington, DC 20546
(202) 453-1000

NASA is the U.S. government agency in charge of the country's civil space program. It publishes a wide variety of brochures, reports, and educational materials—including the quarterly newsletter *Educational Horizons* and *Station Break*, the monthly newsletter covering developments with the space station *Freedom.*

National Space Council
1600 Pennsylvania Ave. NW
Washington, DC 20500
(202) 543-1900

The National Space Council formulates space policy for the United States. It is made up of several cabinet officials and the vice president. It publishes the annual *Report to the President.*

National Space Society (NSS)
922 Pennsylvania Ave. SE
Washington, DC 20003
(202) 543-1900

NSS supports a strong U.S. space program. It is in favor of crewed exploration of space, commercial space activities, and colonization. It publishes the monthly magazine *Ad Astra.*

The Planetary Society
65 N. Catalina Ave.
Pasadena, CA 91106
(818) 793-5100

The Planetary Society, headed by astronomer Carl Sagan, is one of the largest organizations that promotes planetary exploration. It is in favor of both crewed and uncrewed exploration, as well as the search for extraterrestrial intelligence. It publishes a bimonthly magazine, *The Planetary Report.*

United Nations' Committee on the Peaceful Uses of Outer Space (UNCOPUOS)
United Nations
New York, NY 10017
(212) 963-8302

UNCOPUOS is a committee made up of UN member nations that cooperate in peaceful space exploration activities. The committee meets annually to discuss issues related to international cooperation in peaceful space research and uses. It publishes several annual reports, including the *Report on the Peaceful Uses of Outer Space.*

Suggestions for Further Reading

Timothy Levi Biel, *The Challenger*. San Diego: Lucent Books, 1990.

Charles P. Cozic, ed., *Space Exploration*. San Diego: Greenhaven Press, 1992.

Tom D. Crouch, *The National Aeronautics and Space Administration*. New York: Chelsea House Publishers, 1990.

Judy Donnelly & Sydelle Kramer, *Space Junk: Pollution Beyond the Earth*. New York: Morrow Junior Books, 1990.

Jon S. Erickson, *Exploring Earth from Space*. Blue Ridge Summit, PA: TAB Books, 1989.

Barry Evans, *Wrong-Way Comets and Other Mysteries of the Solar System*. Blue Ridge Summit, PA: TAB Books, 1992.

William A. Gutsch, *The Search for Extraterrestrial Life*. New York: Crown Publishers, 1991.

Robert Jastrow, *Journey to the Stars: Space Exploration—Tomorrow and Beyond*. New York: Bantam Books, 1989.

Ben Millspaugh, *Aviation and Space Science Projects*. Blue Ridge Summit, PA: TAB Books, 1991.

Gregory Vogt, *Voyager*. Brookfield, CT: The Millbrook Press, 1991.

Works Consulted

Diane Ainsworth, "Mars Observer: Returning to the Red Planet," *Astronomy*, September 1992.

Edwin Aldrin, *Men from Earth*. New York: Bantam Books, 1991.

Lincoln P. Bloomfield, ed., *Outer Space: Prospects for Man and Society*. Englewood Cliffs, NJ: Prentice-Hall, 1962.

James Carteret, "Time to Rethink Freedom?" *Ad Astra*, September 1991.

Edward Chen, "Big Science Faces Big Troubles," *Los Angeles Times*, June 5, 1991.

Phillip Clark, *The Soviet Manned Space Program*. New York: Orion Books, 1988.

Arthur C. Clarke, *The Promise of Space*. New York: Harper & Row, 1968.

Michael Collins, *Mission to Mars*. New York: Grove Weidenfield, 1990.

Leonard David, "A New Boost for Space?" *Final Frontier*, October 1992.

Bonnie E. Fought, "Legal Aspects of the Commercialization of Space Transport Systems," *High Technology Law Journal*, 1989.

Dale Gardener, "Stepping Out for A Walk," *Final Frontier*, May/June 1992.

Joel K. Harris, "Seeing a Brave New World," *Astronomy*, August 1992.

Richard C. Hoagland, *The Monuments of Mars*. Berkeley, CA: North Atlantic Books, 1987.

Robert Jastrow, *Journey to the Stars: Space Exploration—Tomorrow and Beyond*. New York: Bantam Books, 1989.

James Killian, "Speech to M.I.T. Club of New York, December 13, 1960," *Science*, January 6, 1961.

Michael D. Lemonick, "Echoes of the Big Bang," *Time*, May 4, 1992.

John M. Logsdon, *The Decision to Go to the Moon: Project Apollo and the National Interest*. Cambridge, MA: Massachusetts Institute of Technology, 1970.

Neil McAleer, *The Omni Space Almanac*. New York: World Almanac, 1987.

Thomas R. McDonough, *The Search for Extraterrestrial Intelligence—Listening for Life in the Cosmos*. New York: John Wiley & Sons, 1987.

Robert G. Nichols, "Back to the Future," *Final Frontier*, July/August 1992.

Robert G. Nichols, "Visit to a Red Planet," *Final Frontier*, October 1992.

Judy Pasternak, "Mars Projects Aim to Settle Questions About Red Planet," *Los Angeles Times*, August 20, 1992.

Sally K. Ride, *Leadership and America's Future in Space* (task force report to NASA administrator), August 1987.

Rex Ridenoure, "Doing More with Less: The New Way of Exploring the Solar System," *The Planetary Report*, July/August 1992.

Donald F. Robertson, "Mars on the Cheap," *Final Frontier*, November/December 1991.

Seth Shostak, "Listening for Life," *Astronomy*, October 1992.

Robert W. Stewart and Rudy Abramson, "Experts Struggle to Design Yugo-Priced Space Station," *Los Angeles Times*, May 21, 1993.

Robert W. Stewart and Rudy Abramson, "Imperiled Space Station Puts NASA at Crossroads," *Los Angeles Times*, May 20, 1993.

Robert W. Tinsley, "Nukes to Mars," *Final Frontier*, September/October 1991.

William J. Walter, *Space Age*. New York: Random House, 1992.

Frank White, *The SETI Factor*. New York: Walker & Company, 1990.

John Nobel Wilford, *Mars Beckons*. New York: Alfred A. Knopf, 1990.

Robert Zubrin and Benjamin Adelman, "The Direct Route to Mars," *Final Frontier*, July/August 1992.

Index

About the Author

Anita Louise McCormick has been a free-lance writer since 1988. She specializes in science, nature, and radio-related topics. Her articles and short stories for children, teenagers and adults have appeared in more than three dozen publications nationwide. She is a long-time enthusiast of the space program, a ham radio operator (KA8KGI), and a shortwave radio listener. She recently wrote a book about international radio monitoring—*Shortwave Radio Listening for Beginners*, published by TAB/McGraw-Hill in 1993, and makes regular call-in appearances on the shortwave program "Radio Newyork International."

Picture Credits

Cover photo by The Bettmann Archive

AP/Wide World Photos, 48

Caltech/JPL Facilities, 36

Courtesy of CARA/Caltech, 34

Grafikus, 60, 69

Library of Congress, 12, 59

NASA, 6, 8, 10, 14, 15, 16, 18, 21, 22, 24, 26, 28, 29, 30, 33, 37, 42, 45, 46, 47, 50, 51, 52, 56, 58, 64, 70, 72, 73, 75, 76, 78, 79, 80, 81, 82, 84, 86, 87, 89, 90, 91, 92, 93, 95, 96, 99

Naval Photographic Center, 25

Reuters/Bettmann, 65, 67

Smithsonian Institution, 39

UPI/Bettmann, 13, 20